万达商业规划
销 售 类 物 业

WANDA COMMERCIAL PLANNING 2014
PROPERTIES FOR SALE

2014

万达商业地产设计中心　主编

中国建筑工业出版社

EDITORIAL BOARD MEMBERS
编委会成员

主编单位
万达商业地产设计中心

CHIEF EDITORAL UNIT
Wanda Commercial Estate Design Center

规划总指导
王健林

GENERAL PLANNING DIRECTOR
Wang Jianlin

编委
赖建燕 孙继泉 曲晓东 吕正韬 于修阳 黄国斌 尹强
林树郁 门瑞冰 张东光 曾静 王福魁 陈彬

EDITORIAL BOARD
Lai Jianyan, Sun Jiquan, Qu Xiaodong, Lv Zhengtao, Yu Xiuyang, Huang Guobin, Yin Qiang, Lin Shuyu, Men Ruibing, Zhang Dongguang, Zeng Jing, Wang Fukui, Chen Bin

参编人员
陈文娜 赵龙 刘大伟 孙志超 张天舒 周升森 昌燕
钟山 李金桥 陈晖 李琰 武春雨 黄建好 薛瑜 刘征
薛勇 胡延峰 俞小华 叶啸 荣万斗 文善平

陈海燕 周昳晗 李万顺 李靖 袁文卿

栾赫 武宁 张爱珍 董莉 石亮 赵宁宁 刘敏 高景麟
杨磊 常春林 潘鸿岭 张金菊 刘洋 任睿 邵丽 张博
周恒 刘鹏 纪文青 刘晓敏 彭亚飞 汪新华 申亚男
曹鹏 龚芳 李军 李晅荣 杜文天 车心达 郑德廷 姚韬
漆国强 王嵘 马申申 黄龙梓 刘定涛 郑存耀 周群
张悦 洪剑 李春阳 李子强 周鹏 欧阳国鹏 顾东方
孙一琳 杨健珊 范志满 薛奇 朱作旺 邵强 李鹏
田友军 张克 霍雪影 梁国涛 桑国安 宋波 冯俊 陈涛

PARTICIPANTS
Chen Wenna, Zhao Long, Liu Dawei, Sun Zhichao, Zhang Tianshu, Zhou Shengsen, Chang Yan, Zhong Shan, Li Jinqiao, Chen Hui, Li Yan, Wu Chunyu, Huang Jianhao, Xue Yu, Liu Zheng, Xue Yong, Hu Yanfeng, Yu Xiaohua, Ye Xiao, Rong Wandou, Wen Shanping

Chen Haiyan, Zhou Yihan, Li Wanshun, Li Jing, Yuan Wenqing

Luan He, Wu Ning, Zhang Aizhen, Dong Li, Shi Liang, Zhao Ningning, Liu Min, Gao Jinglin, Yang Lei, Chang Chunlin, Pan Hongling, Zhang Jinju, Liu Yang, Ren Rui, Shao Li, Zhang Bo, Zhou Heng, Liu Peng, Ji Wenqing, Liu Xiaomin, Peng Yafei, Wang Xinhua, Shen Ya'nan, Cao Peng, Gong Fang, Li Jun, Li Xuanrong, Du Wentian, Che Xinda, Zheng Deting, Yao Tao, Qi Guoqiang, Wang Rong, Ma Shenshen, Huang Longzi, Liu Dingtao, Zheng Cunyao, Zhou Qun, Zhang Yue, Hong Jian, Li Chunyang, Li Ziqiang, Zhou Peng, Ouyang Guopeng, Gu Dongfang, Sun Yilin, Yang Jianshan, Fan Zhiman, Xue Qi, Zhu Zuowang, Shao Qiang, Li Peng, Tian Youjun, Zhang Ke, Huo Xueying, Liang Guotao, Sang Guoan, Song Bo, Feng Jun, Chen Tao

校对
陈文娜 陈海燕 周昳晗 李万顺 李靖 袁文卿

PROOFREADERS
Chen Wenna, Chen Haiyan, Zhou Yihan, Li Wanshun, Li Jing, Yuan Wenqing

CONTENTS
目录

A | FOREWORD 012
序言

WANDA PROPERTIES FOR SALE AND DESIGN CENTER 014
万达销售物业与设计中心

B | CULTURAL TOURISM PROJECT 018
文旅项目

VIEWING CULTURAL TOURISM FROM ANOTHER PERSPECTIVE 020
换个角度看文旅

OUTSTANDING PROJECT
优秀项目

01 WUHAN CENTRAL CULTURE DISTRICT WANDA MANSION 024
武汉中央文化区万达公馆

EXHIBITION CENTER
展示中心

01 HARBIN WANDA CITY EXHIBITION CENTER 032
哈尔滨万达城展示中心

02 NANCHANG WANDA CITY EXHIBITION CENTER 036
南昌万达城展示中心

03 HEFEI WANDA CITY EXHIBITION CENTER 042
合肥万达城展示中心

04 QINGDAO ORIENTAL CINEMA EXHIBITION CENTER 046
青岛东方影都展示中心

05 WUXI WANDA CITY EXHIBITION CENTER 050
无锡万达城展示中心

06 GUANGZHOU WANDA CITY EXHIBITION CENTER 056
广州万达城展示中心

PROTOTYPE ROOM
样板间

---

01 PROTOTYPE ROOM OF HARBIN WANDA CITY 062
哈尔滨万达城样板间

02 PROTOTYPE ROOM OF WUHAN CENTRAL CULTURE DISTRICT 064
武汉中央文化区样板间

03 PROTOTYPE ROOM OF NANCHANG WANDA CITY 068
南昌万达城样板间

04 PROTOTYPE ROOM OF WUXI WANDA CITY 072
无锡万达城样板间

05 PROTOTYPE ROOM OF QINGDAO ORIENTAL CINEMA 076
青岛东方影都样板间

06 PROTOTYPE ROOMS OF GUANGZHOU WANDA CITY 080
广州万达城样板间

07 PROTOTYPE ROOMS OF HEFEI WANDA CITY 084
合肥万达城样板间

DEMONSTRATION AREA
实景示范区

01 DEMONSTRATION AREA OF HARBIN WANDA CITY 088
哈尔滨万达城实景示范区

02 DEMONSTRATION AREA OF NANCHANG WANDA CITY 092
南昌万达城实景示范区

03 DEMONSTRATION AREA OF WUXI WANDA CITY 096
无锡万达城实景示范区

04 DEMONSTRATION AREA OF QINGDAO ORIENTAL CINEMA 100
青岛东方影都实景示范区

05 DEMONSTRATION AREA OF HEFEI WANDA CITY 104
合肥万达城实景示范区

COMMERCIAL PROJECT 106
商业项目

ABOUT PROPERTIES FOR SALE OF WANDA COMMERCIAL REAL ESTATE 108
说说万达商业地产的销售物业

OUTSTANDING PROJECT
优秀项目

01 NANJING JIANGNING WANDA MANSION 110
南京江宁万达公馆

02 XI'AN DAMING PALACE WANDA MANSION 116
西安大明宫万达公馆

03 FUQING WANDA PALACE 124
福清万达华府

SALES OFFICE
售楼处

01 SALES OFFICE OF DALIAN JINGKAI WANDA PLAZA 130
大连经开万达广场售楼处

02 SALES OFFICE OF PANJIN WANDA PLAZA 136
盘锦万达广场售楼处

03 SALES OFFICE OF DONGGUAN HUMEN WANDA PLAZA 140
东莞虎门万达广场售楼处

04 SALES OFFICE OF WANDA REIGN CHENGDU 146
成都瑞华酒店售楼处

05 SALES OFFICE OF MEIZHOU WANDA PLAZA 154
梅州万达广场售楼处

06 SALES OFFICE OF SHIYAN WANDA PLAZA 160
十堰万达广场售楼处

07 SALES OFFICE OF NANNING WANDA MALL 164
南宁万达茂售楼处

08 SALES OFFICE OF CHANGSHU WANDA PLAZA 172
常熟万达广场售楼处

PROTOTYPE ROOMS
样板间

01 PROTOTYPE ROOMS OF DONGGUAN HOUJIE WANDA PLAZA 178
东莞厚街万达广场样板间

02 PROTOTYPE ROOMS OF ZHENGZHOU JINSHUI WANDA PLAZA 182
郑州金水万达广场样板间

03 PROTOTYPE ROOMS OF SHENYANG OLYMPIC WANDA PLAZA 186
沈阳奥体万达广场样板间

04 PROTOTYPE ROOMS OF LONGYAN WANDA PLAZA 190
龙岩万达广场样板间

05 PROTOTYPE ROOMS OF SHUDU WANDA PLAZA 194
蜀都万达广场样板间

06 PROTOTYPE ROOM OF SHANGYU WANDA PLAZA 198
上虞万达广场样板间

07 PROTOTYPE ROOMS OF KUNSHAN WANDA PLAZA 202
昆山万达广场样板间

08 PROTOTYPE ROOMS OF CHONGQING BA'NAN WANDA PLAZA 206
重庆巴南万达广场样板间

09 PROTOTYPE ROOM OF DEZHOU WANDA PLAZA 210
德州万达广场样板间

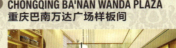

10 PROTOTYPE ROOMS OF YINGKOU WANDA PLAZA 212
营口万达广场样板间

11 PROTOTYPE ROOM OF DONGYING WANDA PLAZA 214
东营万达广场样板间

DEMONSTRATION AREA
实景示范区

01 DEMONSTRATION AREA OF DALIAN HIGH-TECH WANDA MANSION 216
大连高新万达公馆实景示范区

02 DEMONSTRATION AREA OF SHANGHAI JINSHAN WANDA PLAZA 222
上海金山万达广场实景示范区

03 DEMONSTRATION AREA OF SHUDU WANDA PLAZA 226
蜀都万达广场实景示范区

04 DEMONSTRATION AREA OF SIPING WANDA PLAZA 230
四平万达广场实景示范区

EPILOGUE 后续 232

DESIGN CONTROL OF PROPERTIES FOR SALE 234
销售物业设计管控

NEW TASKS OF PROJECT CONTROL OF CULTURE AND TOURISM PROPERTIES FOR SALE 240
文旅销售物业项目管控的新课题

PROJECT INDEX 项目索引 244

FOREWORD
序言

WANDA
COMMERCIAL
PLANNING
2014

WANDA PROPERTIES FOR SALE AND DESIGN CENTER
万达销售物业与设计中心

万达商业地产总裁助理兼设计中心总经理　尹强

万达集团因"商业地产"而闻名，近几年又因文化旅游、金融、影视和体育产业而被广泛关注。虽然销售物业不是万达的核心业态，不被太多人所关注，但每年收获上千亿元现金流的业绩真正支撑着企业的快速发展。经过多年的积累和努力，万达的销售物业已经形成了体系完善、品类丰富、个性鲜明的产品线，在万达进驻开发的大部分城市，都能够成为当地的"销冠"。设计中心作为总部"全面把控销售物业"的管理部门，经过多年的摸索也形成了一套完善的管理体系。

Wanda Group becomes widely known for Commercial Real Estate and catches extensive attention for its cultural tourism, finance, film and television and sports industries in recent years. Although Properties for Sale fails to attract most people's attention since it is not the core business type of Wanda Group, its outstanding achievement harvesting cash flow of hundreds of billion Yuan every year props up rapid development of the enterprise in real sense. Through years of growth and efforts, Wanda Properties for Sale has formed a product line with well-established system, rich categories and distinctive features. In most cities where Wanda has a presence, local "Sales Leader" will undoubtedly falls on it. As an administrative department to comprehensively control Properties for Sale of the headquarter, Design Center, through years of efforts, has fished out a complete set of management system.

一、万达销售物业设计管控——"快与慢"的结合

"快"是万达的显著特点。天下武功，唯快不破。快速开盘、快速销售、快速获取现金流，万达的"快"在业内人尽皆知。在万达，销售物业土地"摘牌"后要求在最短时间内开盘销售，而且销售指标也往往是同城"竞品"项目的数倍。销售指标的达成只简单追求"快"是不行的，在实现快速开发的前提下必须确保产品品质，满足市场定位的需求；否则对速度的追求就会成为无本之木、无源之水。

有"快"就有"慢"，相对于万达开发速度的"快"，前期对于产品设计的推敲是"慢"的。这一方面源自于集团领导对规划设计的重视，另一方面源自于决策过程的审慎和严密。在万达，任何一个项目的产品设计方案都要集团总部审批确认，而且设计方案的审批流程一定是多部门相互确认的。这在很大程度上避免了决策的失误。

万达的"慢"是相对的"慢"。在万达，任何事情都不允许毫无原则地慢慢来。流程的严密是成果品质的保障，绝不能成为延误计划的借口。设计中心作为销售物业规划设计的主管部门，有着一套严密的管理体系，保证了万达的"快"。

1. 产品标准化是"快"的基础

完善的产品标准化体系是实现快速开发的基础。这套产品标准化体系包括从户型标准化、立面效果标准化，乃至夜景照明标准化、精装修标准化、景观部品标准化。

I. DESIGN CONTROL OF WANDA PROPERTIES FOR SALE, A COMBINATION OF "FASTNESS AND SLOWNESS"

Fastness is an outstanding feature of Wanda. There is an old saying goes that fastness is the only martial art that nothing can break. Fast pre-selling, fast selling and fast cash flow earning, fastness has become well known to insiders. In Wanda, properties for sale should start pre-selling within the shortest time after land acquiring, what's more, the sales target is always several times of other "Competition projects" in the city. Of course, pursuit of fastness only can never guarantee achievement of the sales target. On the premise of achieving fast development, the quality must be ensured and demands of market positioning must be satisfied, otherwise the pursuit of speed will be water without a source and a tree without roots.

Where there is fastness, there is slowness. Compared to its fast developing speed, careful consideration on product design at early stage is relatively slow. On one respect, the Group leadership attaches importance to planning and design; for another, the decision-making process itself is prudent and rigorous. As for Wanda, product design scheme of each project should be examined and approved by the Group headquarter and approval process of the design scheme must be a mutually confirmation among several departments, which together avoid decision-making mistakes to a great extent.

Slowness of Wanda is a relative statement, for unprincipled slowness, no matter what, is prohibited in Wanda. Strictness of the process is the guarantee of outcome quality and can never be the excuse of plan delay. The design center, as competent department of Properties for Sale planning and design, has established a strict management system, guaranteeing the Fastness of Wanda.

1. PRODUCT STANDARDIZATION, THE BASIS OF WANDA'S FASTNESS

A complete product standardization system, ranging from standardization of house type and facade effect to nightscape lighting, fine decoration as well as landscape ornaments, is the basis of achieving rapid development.

2. 差异化设计是"快"的要点

每个项目都有着当地市场的特殊需求。简单复制标准化产品是无法满足万达快速销售的要求的，激烈的市场竞争必须通过差异化的产品打开突破口。每个项目都会根据营销定位报告对当地"竞品"项目进行详细的调研，形成每个项目的个性化设计目标；同时，在标准化产品模块中，特别设置了差异化产品设计模块，便于项目灵活选用，形成有竞争力的差异化产品。

3. 严密的管理是"快"的保障

对于方案设计成果有严密的管理流程，通过明确的审核要点多级审核，确保产品品质。从项目评审、设计中心专业总工审核，到设计中心领导审核、营销成本跨部门审核，乃至集团主管领导审批，每个项目都要经过这样一系列分工严密的审核、审批。

设计中心"重点把控效果类方案"，负责标准化设计、产品研发和管理工具的制定；项目公司负责施工图设计、工艺深化图纸把关和现场实施。针对不同的职责，对应各层级的考核标准，均建立一整套行之有效的考核体系，为实现"快"的目标保驾护航。

二、万达销售物业设计管控——"前与后"的衔接

前期的规划设计通过"快"与"慢"的有机结合，确保了设计成果的高品质。高品质的设计成果如何能够完美地实现，这就需要"前"与"后"的紧密衔接。

在万达，设计中心要对项目规划设计的全过程负责。但事有"轻重缓急"之分，设计中心需要合理调配人力资源和时间成本，分清主要问题和次要问题、主要环节和次要环节、主要矛盾和次要矛盾，采取分级管理的原则。前期抓设计亮点，后期抓管控流程，通过"前"与"后"的紧密衔接，确保设计成果完美"落地"。

1. 调动设计资源

建立第三方审查制度、引入第三方设计单位，对施工图设计质量进行审查把关。施工图审查得分计入对设计单位的履约评估。通过建立一整套针对设计成果质量的考核机制，引导设计单位进行正向竞争。

2. 优化管理工具

在设计中心，对于管理工具的优化与完善一直持续

2. DIFFERENTIATION DESIGN, THE KEY POINT OF WANDA'S FASTNESS

Obviously, each project has its own special needs due to local markets. Simply copying standardized products can't satisfy the requirements of quick selling, and only differentiated product can make breakthrough facing fierce market competition. Each project will, based on the marketing positioning report, be subject to detailed investigation and survey on local competition projects, in an attempt to form a distinctive design objective for each project. Meanwhile, differentiated product design module is specially set in the standardized product module for flexible use and helps to form competitive differentiated product.

3. STRICT MANAGEMENT, THE GUARANTEE OF WANDA'S FASTNESS

Strict management process is applied to scheme design outcomes with clear checking points set for multi-level approval, thus to guarantee the product quality. Every product must accept a series of checking and approval in strict division, including project review, approval by specialty chief engineer of the design center, by leadership of the design center, by cross-department for marketing cost and by competent leader of the Group.

The design center put key emphasizes on controlling rendering scheme and setting standardized design, product R&D and management tool. The project company takes charge of construction drawing design, process development drawing checking and field implementation. As for different duties and corresponding different levels of evaluation standard, a complete set of effective evaluation system is set to escort the goal of Fastness.

II. DESIGN CONTROL OF WANDA PROPERTIES FOR SALE, AN INTEGRATION OF "EARLY AND LATER STAGES"

Planning design at the early stage, benefiting from the dynamic integration of "Fastness" and "Slowness", guarantees high quality of the design outcome, the perfect realization of which, however, requires tight connection of Priority and Postponement.

In Wanda, the design center should be responsible for the whole project planning and design process. Things have priorities, therefore, the design center should, under the principle of level-to-level administration, reasonably allocate human resources and time cost, and draw a clear distinction between the primary and secondary issues, links and contradictions. Through putting emphasis on design highlights at the early stage and pay attention to management and control process at the later stage, and a tight integration of Priority and Postponement, design outcomes are surely to be established smoothly and perfectly.

1. MOBILIZE DESIGN RECOURSES

Establish the third party review system and introduce the third party design unit, in an attempt to examine and control quality of the construction drawing design. Score of the construction drawing review will be included in the performance evaluation of the design unit. Wanda, through establishing a complete set of evaluation mechanism on quality of design outcomes, guides design units to a positive competition.

2. OPTIMIZE MANAGEMENT TOOL

进行从未间断。从无到有、从有到全、从全到精、从精到合，将多年的管控经验进行总结提炼汇编成《操作手册》。这种"升华"，使得管控工具简明有效、管理指令清晰、品质标准明确可视，推行之后现场成果品质逐年提升。

3. 完善管控流程

为保证前期工作与后续的执行紧密衔接，设置了严密有效的管控流程——从方案交底会、成果论证会到第三方审查成果复审会——对各阶段的设计成果都有明确的审核要点和严谨的管理流程。同时，推行项目安全排查制度，重视过程管理，鼓励项目公司采取"自查"的方式，使得问题早发现、早解决，防患于未然。

三、万达销售物业设计管控——"上与下"的协作

"前"与"后"的紧密衔接，离不开"上"与"下"的分工协作。因为有了分工协作，集团与项目公司形成了"管控共同体"。集团不断研发、提升产品竞争力，项目公司利用管控工具严格执行，确保实施品质。

"上"与"下"之间不是截然分开，而是相互理解与互动。设计中心组织各类产品、管理研讨会，利用出差"下"项目培训、"上交"出差报告、分析、总结现场管控经验等心得和信息，将集团的要求和标准、项目的困难与经验"上传、下达"，为统一工作目标起到重要的作用。

设计中心的"星火计划"，将一百多个项目公司分为17个区域；每个区域设立各专业带头人，通过"上下"交流、"横向"学习，激发基层员工的学习积极性。通过交流，发现优秀人才调动到设计中心或重点项目定向培养；同时设计中心也有许多优秀的管理人才到项目公司得到锻炼和发展。

四、万达销售物业设计管控——"新与旧"的更替

发展至今，万达集团已经历了"跨区域、商业地产、文化旅游、跨国发展"等阶段，可谓"常变常新"。伴随着集团的发展，设计中心和销售物业的管理人员也因为新需求和高要求而逐步更新，"新人新事"、"旧人旧事"有序更迭，快速转换，保证了万达集团的发展速度。

As for the design center, optimization and perfection of management tool is a continuous task. Developing from nothing, from incompleteness to completeness, from completeness to expertness, from expertness to integration, the design center sums years of experiences on control and management up and compiles an Operation Manual. Such enhancement makes the management and control tool concise and effective, the executive instruction clear and the quality standard clear and definite. Benefiting from application of such enhancement, quality of the filed outcome is improved year by year accordingly.

3. COMPLETE MANAGEMENT AND CONTROL PROCESS

To guarantee tight connection between preliminary work and follow-up implementation, a well-conceived and effective management and control process covering scheme disclosure, outcome discussion meeting and outcome review meeting by the third party is established, specifying clear review points and rigorous management process targeting design outcomes at each stage. What's more, project security investigation policy is applied, process control is emphasized and self-inspection mode is encouraged to be adopted by the project company, in an attempt to achieve early detection and solving of hidden problems and prevent negative consequences before they emerge.

III. DESIGN CONTROL OF WANDA PROPERTIES FOR SALE, A COOPERATION OF "HIGHER AND LOWER LEVELS"

The tight integration of "early and later stages" is hard to achieve without the cooperation of "higher and lower levels", benefiting from which the "Management and Control Community" of the Group and project company can be formed. The former focuses on researching and improving product competitiveness continually, and the latter, by management and control tools, concentrates on strict execution to guarantee implementation quality.

The "higher and lower levels" are a kind of mutual understanding and interaction but are not completely separated. The design center takes charge of organizing all kinds of products and management seminars, issuing project training through missions, submitting reports and analysis on the business trip, summarizing information and data concerning field management and control experiences, reporting and transmitting requirements and standards, as well as difficulties facing the project and experience gaining from the project, thus making an important role in setting an unified objective.

"Spark Program" of the design center divides hundreds of project companies into seventeen regions with each region appointed with specialty leader, which, through communication at "higher and lower levels" and "horizontal" learning, motivates initiatives of grass-roots employees. Through communication, talents spotted are to be transferred to the design center or key projects for orientation training, meanwhile, excellent management talents will be transferred to project companies for training and development.

IV. DESIGN CONTROL OF WANDA PROPERTIES FOR SALE, A REPLACEMENT OF "THE OLD BY THE NEW"

"新与旧"之间存在必然的关联。万达的创新发展都是在总结的基础上创新,把"旧事"做成"新事";做"旧事"也有新要求,市场在变,设计更要变。

企业发展的核心是"利润"。销售物业是万达的"造血"机器,设计中心是这部机器的"中枢"。2014年是设计中心的创新年,面对集团新格局和新模式,设计中心积极开拓新思维,研发新的设计产品支持营销,制定新的管控工具支持集团发展。

2015年,万达集团将迎来新的转型,设计中心也将面临新的挑战,让我们认真总结、沉淀过往,潜心规划、拥抱未来!

Till now, Wanda Group, experiencing diverse stages including cross-regional, commercial real estate, cultural tourism and multinational development, keeps changing. Along with development of the Group, administrative staffs of the design center and properties for sale keep continuous development motivated by new needs and higher requirements. Orderly alternation and rapid conversion of "new people & new deeds" and "old people & old deeds" guarantee development speed of Wanda Group.

"The new and the old" must have a kind of necessary linkage. Innovation and development of Wanda Group are always based on summarization of the past, making the "old deeds" to be "new deeds". Of course, new requirements are inevitable for doing the old deeds as the changing market appeals to changes on design.

The core of enterprise development falls on the "Profit". As for Wanda Group, properties for sale is the hematopoietic unit, for which design center is the main center. In 2014, the year of creativity and innovation for the design center, the design center, facing new pattern and model of the Group, develops new thinking actively, researches and develops new design products supporting marketing and establishes new management and control tools for backing Group development.

In 2015, Wanda Group will witness a new transition, and the design center will be faced with new challenges. Let's summarize carefully, retrospect the past, plan with great concentration and embrace the future.

CULTURAL TOURISM PROJECT
文旅项目

WANDA
COMMERCIAL
PLANNING
2014

VIEWING CULTURAL TOURISM FROM ANOTHER PERSPECTIVE
换个角度看文旅

万达商业地产设计中心副总经理兼文旅设计部总经理　门瑞冰

在"外界"眼中，万达文化旅游城（万达城）给人们最直观、最深刻的印象是"规模巨大"。除了大，文旅项目还具有业态丰富、建筑类型复杂、开发周期漫长、品质要求极高等诸多特点，在"万达人"自己"口中"，常念叨的是各个万达城的数据、计划进度和品质提升方法。

由于工作的原因，作为万达商业地产文旅销售物业的设计管理者，我们习惯于站在开发商的角度，用设计师或者设计管理者的视角去审视万达城；也可能忽视一个问题：真正与万达城朝夕相处，甚至可能是相伴一生的客户、业主，他们眼中的万达城有什么特点？其实，他们的看法更具说服力、更具代表性，也是我们更应该关注的。

在这里，我想尝试站在一个客户、一个业主的角度，去重新"发现"万达城的特点——换个角度看文旅，会有新的发现，也会为我们未来万达城的设计管控工作带来新的启示。

客户眼中，万达城第一个特点："高品质"

品质，在任何时候、任何项目中，毫无疑问都是客户关注的第一要素。

万达城是超大规模的城市综合体，复合了多种功能业态，为客户打造的是以"改善类"居住、生活体验为主的复合社区。而"改善类"客户，已经不再单纯地关注产品的户型、面积、价格，进而开始关注环境、品质、圈层、升值潜力等项目的"软实力"。他们对万达城销售物业的品质要求比一般项目更高，除了常规意义上的建筑结构、园林景观等硬件品质外，客户还要求生活感受、交通便捷、配套服务、文化底蕴等软件品

In the eyes of "outsiders", the most intuitive, the deepest impression aroused by Wanda Cultural Tourism City (Hereinafter referred to as Wanda City) is its "large scale". Except from large scale, Wanda City is inclusive of varied features, such as enriched business types, complex building type, lengthy development cycle and high quality requirement, etc.; however, in the "mouths" of "Wanda Staff", the data, schedule and quality promotion methods of the Wanda City are repeatedly talked about.

Due to work commitments, the design managers of Cultural and Tourism Properties for Sale of Wanda Commercial Real Estate are used to standing in the angle of the developer to judge Wanda City from the perspective of designer or design manager. In the process as such, one issue maybe ignored comes out: What's the feature of Wanda City in the eyes of clients and owners who are closely associated with or even lifelong companion of Wanda City in the real sense? Their opinions, in fact, are more convincing, representative and concerned.

Hence, I hereby strive to rediscover the features of Wanda City from the standpoint of a client and an owner. Judging the cultural and tourism city from a new perspective is to bring fresh discoveries and enlightenment to the future design control of Wanda City.

IN THE EYES OF THE CLIENT, "HIGH QUALITY" IS THE FIRST FEATURE OF WANDA CITY

Undoubtedly, quality is the No.1 concern of client at any time for any project.

As a super-large scale urban complex, Wanda City integrates a variety of functional formats and builds a community complex that highlights improved housing and life experience. While the clients demanding improved housing, instead of simply focusing on house type, area and price of products, show concern to the projects' "soft power" such as environment, quality, circle layer, appreciation potential. Their quality requirements on Wanda City Properties for Sale are more demanding compared with general projects, as they, in addition to the conventional

（图1）广州万达城总平面图

（图2）广州万达城展示中心

质。而这类"软实力、软品质"的打造，对于建筑硬件品质的打造更为严苛，要求也更高。

另外，万达城较长的建设周期决定了客户对万达城品质要求更高。一个万达城，仅销售物业部分，总建筑面积动辄二三百万平方米，甚至达到四五百万平方米。超大的建筑规模，意味着开发过程是一个较长的周期，需要分期、分批开发。和所有消费品一样，客户对分期开发的销售物业也会有一个不断升级换代、品质提升的诉求，总是认为后期的产品品质应该超越前期的产品。这就对万达城销售物业的品质提出了不断提升的要求，对我们设计管控的工作提出了更高的要求。这就是万达城第一个突出特点——"高品质"。

客户眼中，万达城第二个特点："乐生活"

万达城能够为客户提供丰富多彩的生活体验。

万达城不同于传统的纯居住郊区大盘，除了住宅和底商之外，万达城还复合了万达茂、主题公园、秀场、酒店及酒吧街等一系列旅游、休闲功能业态，使其由单一的居住区，升级为集旅游、休闲、居住等功能为一体的综合体。

多功能复合的业态，让人们可以有机会去体验更加丰富多彩的快乐生活——从晨起到入睡、从春天到冬天、从小孩到大人和老人、从吃饭到休闲、从看电影到雪上运动——让各色人等在万达城各取所需。这就是万达城第二个突出特点——"乐生活"。

客户眼中，万达城第三个特点："朋友圈"

万达城可以为客户维系、强化亲情、友情、联系人脉提供可能。

我们都有过搬家的经历。但是无论何种原因，每次搬家，我们都会失去一些东西，比如闲置的家具电器、淘汰的旧衣物、过期的杂志报纸……这些东西随着时间推移还会慢慢再积累起来。但有一样东西却是搬家后很难再寻回的，那就是依附于原有居住地的"朋友圈"。

在城市中，由于居住小区规模相对较小、产品类型相对单一，搬家往往意味着要更换社区，直接导致原有地缘人脉关系的断裂。距离的疏远是导致搬家成为"朋友圈"杀手的关键原因。而万达城由于规模巨大、产品类型丰富多样，在一个万达城内，涵盖可以满足家庭不同生命周期居住需求的各类产品——

PART B — CULTURAL TOURISM PROJECT — 文旅项目

hardware quality of building structure and garden landscape, appeal to the software quality of life experience, convenient transportation, supporting service, cultural deposit, etc. Yet compared with architectural hardware, the construction of "soft power and soft quality" seems to be more rigorous and demanding.

The lengthy construction cycle of Wanda City also contributes to this higher request of client on quality of Wanda City. As for Wanda City, the properties for sale only is to generally cover a total floor area of 2 to 3 million square meters, even 4-5 million square meters. The oversized construction scale naturally leads to a longer development cycle requiring phased and split development. Likewise, like all other consumer goods, clients have upgraded and quality-improving appeal against the properties for sale in phased development, always expecting more excellent products in later phase that that in earlier phase. This proposes constantly improving requirement on quality of properties for sale of Wanda City and more demanding requirements on design control work. The first prominent feature of Wanda City is thus presented-"High Quality".

IN THE EYES OF THE CLIENT, "HAPPY LIFE" IS THE SECOND FEATURE OF WANDA CITY

Wanda City is capable of providing clients with colorful life experience.

Unlike the traditionally pure residential suburban project, Wanda City combines a series of tourism and leisure functional formats such as Wanda Mall, theme park, show theatre, hotel and Bar Street, etc. besides residence and commerce at the bottom, upgrading it to a complex that integrates tourism, leisure, residence, and other functions from a single residential area.

Multi-functional composite business types offer people the chance to experience more colorful and happier life, acquiring anything they want in Wanda City from the dawn to night, from spring to winter, from kids to adults and the elderly, from eating to leisure and from movies to skiing sports. This constitutes the second prominent feature of Wanda City- Happy Life".

IN THE EYES OF THE CLIENT, "FRIEND CIRCLE" IS THE THIRD FEATURE OF WANDA CITY

Wanda City may help clients to maintain and enhance family ties, friendship and connections.

Whatever the reason, in our each ever-experienced moving process, we are sure to lose something, such as idle furniture and electrical appliances, used clothes, expired magazine and newspaper... All these things are possible to be slowly accumulated over time, yet one thing can hardly recover after the move-the "Friend Circle" that is attached to the original residence.

Due to the relatively small size, relatively monotonous product type of urban residential areas, move often means changing community and directly leads to the difficulty of maintaining original geopolitical connections. It is the distance that largely makes move become the killer of "friend circle". While a Wanda City, boasting of huge scale and diverse product types, encompasses various products that cater for housing needs of families at different life cycle, which means that a family can realize full-life cycle residence in Wanda City and

也就意味着一个家庭,可以在万达城内完成全生命周期的居住生活,这就为维系"朋友圈"提供了可能。

客户眼中,万达城第四个特点:"多文化"

每一个万达城,都有其独特的文化特质。

不同于普通城市居住区或是一般郊区大盘的文化"舶来化",万达城的文化特质是有根有据、有生命力的。它既是由万达城所处城市、区域的历史、地理环境所赋予的,也是万达城从规划、产品,到空间、立面设计过程中着力打造的,是对其所在城市或区域文化的延续与升华。例如西双版纳万达城的澜沧江民族风情文化、哈尔滨万达城的俄罗斯风情文化、青岛维多利亚湾的海洋文化等。

万达城的文化特质,既是影响客户选择万达城的重要决策因素,也是客户在万达城的生活中建立归属感、认同感的重要因素。强烈的文化特征,会赋予万达城鲜明的城市、社区个性;而鲜明的个性,更容易吸引具有共同认知、喜好、背景的客户,更容易在较短时间内建立起居民共同认可的城市、社区的生活、文化准则,从而促进居民对万达城建立认同感和归属感。这就是万达城第四个突出特点——"多文化"。

makes it possible to keep "Friend Circle".

IN THE EYES OF THE CLIENT, "MULTI-CULTURE" IS THE FORTH FEATURE OF WANDA CITY

Each Wanda City is endowed with its unique cultural trait.

Different from the "imported" culture available in ordinary urban residential area or the average suburban project, Wanda City's cultural trait are well-founded and have vitality. It is given by the historic and geographical environment of the city and region where the Wanda City locates, and consciously built from planning, product to space and facade design of Wanda City. The cultural trait is the continuation and sublimation of the city or the region concerned, such as the Lancang River folk customs and culture in Xishuangbanna Wanda Cultural Tourism City, Russian customs and culture in Harbin Wanda Cultural Tourism City, and ocean culture in the Qingdao Victoria Bay, etc.

Cultural trait of Wanda City is served as both a critical decision factor to affect clients' selection of Wanda City and a key factor for clients to establish a sense of belonging and identity when living in Wanda City. Strong cultural trait endows Wanda City with distinguishing city and community uniqueness, which in turn helps to attract clients with a common cognition, tastes and backgrounds, establish life and cultural norms of city and community winning residents' mutual recognition in a short time, and eventually build a sense of belonging and identity toward Wanda City among residents. This constructs the fourth prominent feature of Wanda City-"Multi-Culture".

(图3)南昌万达城展示中心

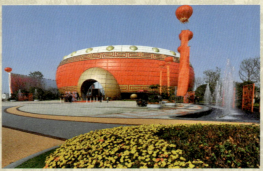

(图4)合肥万达城展示中心

(图5)无锡万达城展示中心

客户眼中，万达城第五个特点："理想"

万达城为客户实现个人的理想提供了可能。

超大的规模、较长的建设周期、巨量的人口、复合的功能、丰富的业态、美好的前景，使每个人都有机会在其中找到自己想要的东西。无论是对生活状态和品质的诉求，还是对创业梦想的实践，或是对业余爱好的追求，在万达城中，都有其实现的空间和可能性。这就是万达城第五个突出特点——"理想"。

万达文旅城，无论是在我们万达人眼中，还是在客户、业主眼中，或许看到的内容、特点各不相同，愿意为其付出和希望从中得到的东西也会千差万别；但是有一点是相同的，那就是希望它能够健康成长；随着时间的延续，能够成为我们共同认可的理想家园。

我们希望通过我们的努力，能够让万达城成为可以满足人们一天之中生活工作、休闲娱乐等各种需求的"一天之城"；希望万达城能够成为满足人们全生命周期各种需求，成为让人们愿意与之相伴一生的"一生之城"！

IN THE EYES OF THE CLIENT, "IDEAL" IS THE FIFTH FEATURE OF WANDA CITY

Wanda City makes it possible to realize personal ideal of the clients.

Oversized scale, lengthy construction cycle, massive population, composite functions, rich business types and rosy prospects give everyone a chance to spot what they want. In Wanda City, you are left with the room and possibility to realize the appeal to living conditions and quality, practice of entrepreneurial dream, or the pursuit of a hobby. This is the fifth prominent feature of Wanda City-"Ideal".

From the perspectives of Wanda Staff and client and owner, Wanda Cultural Tourism City may seem different in terms of its content and feature and what they pay for and want to get from it also differ a lot, yet the wish for its healthy growth to become our mutually recognized ideal home as time goes by makes no different.

We wish, through our efforts, to make Wanda City the "Daily City" that can satisfy diverse daily demands of people ranging from life, work, leisure and recreation, and the "Lifelong City" that can meet full-life cycle demands of people and accompany people for life long.

2014 WANDA COMMERCIAL PLANNING
万达商业规划——销售类物业

优秀项目 01

WUHAN CENTRAL CULTURE DISTRICT WANDA MANSION
武汉中央文化区万达公馆

项目位置 湖北/武汉	**LOCATION** WUHAN / HUBEI PROVINCE
占地面积 9.75 公顷	**LAND AREA** 9.75 HECTARES
建筑面积 66.13万平方米	**FLOOR AREA** 661,300m²

PROJECT OVERVIEW
项目概况

武汉中央文化区位于武汉市武昌区，规划区域1.8平方公里，总建筑面积340万平方米，由汉街、万达广场、汉秀、电影乐园、七星级酒店、五星级酒店及高端写字楼等业态组成，集文化艺术、旅游休闲、商业娱乐、商务办公和高端居住五大功能于一体。万达公馆为中央文化区的顶级精品住宅。

Located in Wuchang District, Wuhan City, Wuhan Central Culture District, with a planned area of 1.8 square kilometers and a total floor area of 3.4 million square meters, consists of Han Street, Wanda Plaza, Han Show, movie park, seven-star hotel, five-star hotel, high-end office building and other types of business, and integrates five functions (i.e. culture and art, tourism and leisure, business and entertainment, business office and high-end residence) into one. Wanda Mansion constitutes its top quality residence.

01

02

PART B | CULTURAL TOURISM PROJECT
文旅项目

01 武汉中央文化区万达公馆建筑外立面
02 武汉中央文化区万达公馆总平面图
03 武汉中央文化区万达公馆夜景

04

05

04 武汉中央文化区万达公馆入口
05 武汉中央文化区万达公馆建筑外立面
06 武汉中央文化区万达公馆夜景

PLANNING DESIGN
建筑规划

整体规划布局开阔、方正，建筑栋距达到百米，欧式古典园林和ART DECO建筑风格彰显豪宅品质。高耸的住宅建筑群北临沙湖，可俯瞰繁华的楚河汉街，将沙湖、水果湖、东湖三大著名景观湖相连，成为武汉独一无二、不可复制的一线观湖豪宅。

建筑采用"三段式"、对称的立面构图原则，选用深棕色干挂石材、结合米灰色高级仿石涂料，饰以精美雕刻，体现对品质的极致追求。顶部层层收缩呈阶梯状，强调高耸的垂直线条，形成类金字塔的效果，使建筑整体高挑挺拔！

Enjoying a spacious and upright general planning layout, the towering residential buildings boast of a-hundred-meter clearance and European classical garden and ART DECO architectural styles that reveal the quality of a mansion, and overlook the prosperous Chu River and Han Street as the buildings face Shahu in the north. Linking three famous landscape lakes of Shahu, Shuiguohu and Donghu together, Wanda Mansion grows to be a unique luxury lake view residence in Wuhan.

Following "three-section" and symmetrical facade composition principle, the mansion chooses dark brown dry hanging stone combined with beige gray advanced stone like paint and decorated with exquisite carving to show its ultimate pursuit of quality. Besides, the ladder like layered contraction at the top highlights the towering vertical lines and forms a Pyramid like effect, presenting a tall and straight architectural image.

LANDSCAPE DESIGN
景观

公馆景观为欧式古典主义园林风格，与建筑风格相匹配，强调入口轴线景观，将各个不同功能、形态的区域连接成紧凑而丰富的序列空间。借鉴中国传统园林中对景、夹景、借景等手法，使园林小品、植物、建筑交相呼应。利用地形高差，建造错落有致的精致水景，营造一派高贵典雅的园林氛围。

Echoing the architectural style, the landscape of Wanda Mansion presents European classical garden style. With emphasis on the entrance axis landscape, it connects areas with diverse functions and forms into a compact and abundant sequence space; borrowing techniques of opposite scenery, vista and view borrowing from Chinese traditional gardens, it enables garden ornaments, plants and buildings to eco each other; taking advantage of terrain elevation difference, it constructs well-spaced delicate waterscape. A Noble and elegant garden atmosphere is thus felt.

07 武汉中央文化区万达公馆水景
08 武汉中央文化区万达公馆景观
09 武汉中央文化区万达公馆入口喷泉
10 武汉中央文化区万达公馆雕塑

B | CULTURAL TOURISM PROJECT
文旅项目

09

10

INTERIOR DESIGN
内装

室内精装定位为法式新古典主义风格。淡雅的卡布其诺色调，搭配"点睛"的金色配饰，在柔和的灯光下静静地释放出独特的浪漫气息。天然大理石的自然纹理、精雕细琢的装饰线条与华丽的手绘壁纸相呼应，搭配手工定制家具饰品，营造18世纪法式宫廷古典空间。

Interior fitting-out pursues French neoclassical style. The simple and elegant cappuccino color going with focused gold accessories quietly creates a unique romantic atmosphere in the soft light. The harmony between natural texture of natural marble, exquisite decorative lines and magnificent hand-painted wallpaper, and the display of hand-tailored furniture accessories together build the 18th century French court classic space.

11 武汉中央文化区万达公馆客厅内装
12 武汉中央文化区万达公馆门厅内装
13 武汉中央文化区万达公馆餐厅内装
14 武汉中央文化区万达公馆卧室内装

PART B CULTURAL TOURISM PROJECT
文旅项目

12

13

14

2014 WANDA COMMERCIAL PLANNING
万达商业规划——销售类物业

HARBIN WANDA CITY EXHIBITION CENTER
哈尔滨万达城展示中心

开放时间 2013 / 07 / 20	**OPENED ON** JULY 20 / 2013
项目位置 黑龙江 / 哈尔滨	**LOCATION** HARBIN / HEILONGJIANG PROVINCE
占地面积 1.76 公顷	**LAND AREA** 1.76 HECTARES
建筑面积 0.40 万平方米	**FLOOR AREA** 4,000m²

BUILDING OF EXHIBITION CENTER
展示中心建筑

哈尔滨万达城展示中心造型宛如一个"冰壶"，通过细致的方案设计和材料选择，使其在完美地展现"冰壶"形象的同时，满足了"文化旅游城"销售、展示等使用功能的要求；既与哈尔滨"冰雪之城"的特点相契合，也与"万达茂冰雪大世界"业态相呼应。

Through careful scheme design and material selection, "curling" shaped Harbin Wanda City Exhibition Center, while perfectly displaying the "curling" image, caters for use function requirements on sale and exhibition as a Cultural Tourism City. The building both corresponds to the features of "the ice city" and echoes with "Wanda Mall Ice & Snow World" format.

01 哈尔滨万达城展示中心总平面图
02 哈尔滨万达城展示中心鸟瞰图
03 哈尔滨万达城展示中心建筑外立面

PART B　CULTURAL TOURISM PROJECT
文旅项目

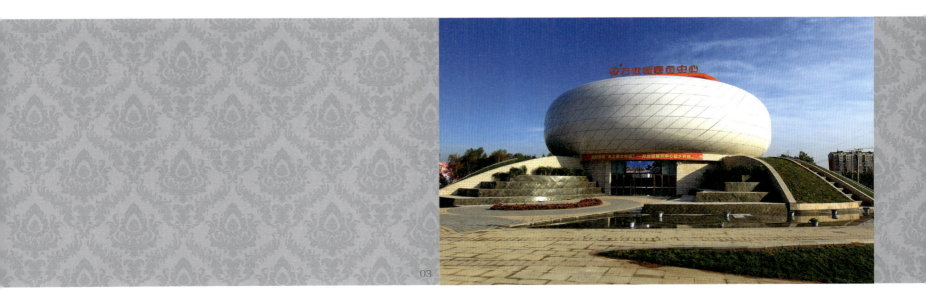

03

PART B CULTURAL TOURISM PROJECT
文旅项目

INTERIOR OF EXHIBITION CENTER
展示中心内装

内装设计主题取材于"一片冰心在玉壶"。平面功能围绕中心沙盘展开——休息区、水吧区、VIP区及LED屏幕环绕其周围,不仅在功能上满足了需求,在平面形态上也延续了建筑"冰壶"的感觉,做到内外合一、相辅相成。互动触摸屏、观影区及3D体验厅,可以让客户亲身体验冰壶运动的无穷魅力,尤其是3D体验厅如同巨幕一般的LED屏,提供了令人震撼的视觉体验。

Sourcing its theme from "chasteness of soul", the interior design has its plane function (rest area, water bar area, VIP area and LED screen) spread by focusing on central building model, which satisfies functional requirements and inherits the architectural curling sense in terms of plane form, and realizes inside & outside combination and complementary relation. Interactive touch screen, viewing area and 3D experience hall offer customers opportunity to experience the endless charm of curling. The 3D experience hall in particular, with a giant LED screen, provides a stunning visual experience.

04 哈尔滨万达城展示中心大厅内装

2014 WANDA COMMERCIAL PLANNING
万达商业规划——销售类物业

展示中心 02

NANCHANG WANDA CITY EXHIBITION CENTER
南昌万达城展示中心

开放时间 2013 / 08 / 18	OPENED ON AUGUST 18 / 2013
项目位置 江西 / 南昌	LOCATION NANCHANG / JIANGXI PROVINCE
占地面积 2.1 公顷	LAND AREA 2.1 HECTARES
建筑面积 0.53 万平方米	FLOOR AREA 5,300m²

01 南昌万达城展示中心建筑外立面
02 南昌万达城展示中心总平面图

01

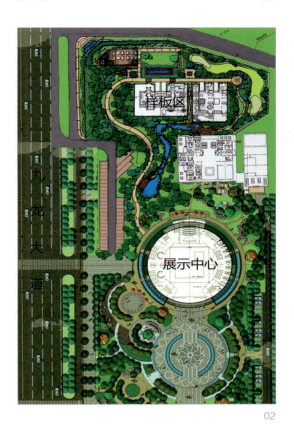

BUILDING OF EXHIBITION CENTER
展示中心建筑

南昌万达城展示中心采用极具民族特色的"青花瓷"来表现。由广场方向望去，首先映入眼帘的是三片"中国红"飘带构成的入口，仿佛正迎接远道而来的客人。"中国红"与"青花瓷"产生的颜色对比增强了入口的标识性，也增强了空间的序列感。建筑立面采用夹宣玻璃模拟青花瓷"白如玉"的效果，散发出简洁、纯净、自然的气质。外表面的牡丹纹饰代表着"花开富贵"，象征着繁荣昌盛、富贵和平。

Nanchang Warda City Exhibition Center is presented with "blue and white porcelain", full of national features. Seen from the plaza, the entrance made of three "China red" ribbons firstly jumps into our sight, looking as if it is greeting the distinguished guests afar. By virtue of the color contrast of "China red" and "blue and white porcelain", both the identity of the entrance and sequence sense of space are enhanced. In terms of the building facade, silk laminated glass is employed to imitate the jade white effect of blue and white porcelain, revealing a concise, pure and natural temperament. When it comes to the external surface, the peony pattern represents the Fortune coming with blooming flowers, a symbol of prosperity, wealth and peace.

CULTURAL TOURISM PROJECT
文旅项目

03 南昌万达城展示中心景观
04 南昌万达城展示中心夜景
05 南昌万达城展示中心水景
06 南昌万达城展示中心景观廊架
07 南昌万达城展示中心景观

LANDSCAPE OF EXHIBITION CENTER
展示中心景观

秉承"自然、生态、健康、休闲"的设计理念，结合"青花瓷"外立面效果，展示中心采用了现代中式园林景观风格，由中轴大广场、水景、廊架、木平台、景框、景石、植栽、花钵及小品等形成张弛有度的空间序列，整体布局统一，景观元素变化多，客户体验丰富。

Adhering to the design concept of "Nature, Ecology, Health and Leisure", and integrating the "Blue and White Porcelain" facade effect, the exhibition center adopts the modern Chinese garden landscape style. Arranging the middle axis grand plaza, waterscape, gallery, wooden platform, scenic frame, scenic stone, planting, flower pot and featured landscape to form a flexible space sequence, the design achieves unified layout as a whole, varied landscape elements and rich customer experience.

08 南昌万达城展示中心接待台
09 南昌万达城展示中心大厅内装
10 南昌万达城展示中心沙盘

INTERIOR OF EXHIBITION CENTER
展示中心内装

室内设计采用了中轴对称、大尺度的环形中庭，通过直径18米的彩色玻璃发光顶、中国传统"椽头"造型灯槽与玉石"青花瓷"发光地面，达到"上下呼应"的效果，完美地表现了"青花瓷"的神韵，表达了"金盘"的寓意。中庭四周8根饰有白色"祥云"图案的"中国红"钢柱以及刻有"中国印"的白色饰面板，使展示中心室内空间清雅中透露出热烈，满足营销展示的功能需求。

In the large scale axisymmetric circular atrium, the interior design, through the adoption of stained glass roof with diameter of 13m, Chinese traditional sally-shaped light trough and jade "Blue and White Porcelain" luminous ground, attains the upper-lower harmony, perfectly shows the charm of "Blue and White Porcelain" and expresses the moral of "Gold Plate". In addition, eight "China Red" steel columns emblazoned with white "Auspicious Clouds" surrounding the atrium and white panels engraved with the "Chinese Seal" forge an elegant while ardent exhibition center interior, satisfying the functional requirement for marketing exhibition.

2014 WANDA COMMERCIAL PLANNING
万达商业规划——销售类物业

HEFEI WANDA CITY EXHIBITION CENTER
合肥万达城展示中心

开放时间 2013 / 11 / 02	**OPENED ON** NOVEMBER 2 / 2013
项目位置 安徽 / 合肥	**LOCATION** HEFEI / ANHUI PROVINCE
占地面积 1.15公顷	**LAND AREA** 1.15 HECTARES
建筑面积 0.47万平方米	**FLOOR AREA** 4,700m²

BUILDING OF EXHIBITION CENTER
展示中心建筑

建筑设计借鉴凤阳花鼓的肌理和形态，寓意着喜庆祥和。建筑整体高18米，直径61米，建筑面积达4700平方米，2014年荣获"大世界吉尼斯之最"——世界最大单体鼓形建筑。

Imitating the texture and morphology of Fengyang Flower Drum Dance, the architectural design signifies joy and peace. With overall height of 18 meters, diameter of 61 meters and floor area of 4700 square meters, in 2014, Hefei Wanda Exhibition Center won the "Guinness Record" of "World's Largest Single Drum-Shape Architecture".

01

02

PART B CULTURAL TOURISM PROJECT
文旅项目

01 合肥万达城展示中心总平面图
02 合肥万达城展示中心建筑外立面

LANDSCAPE OF EXHIBITION CENTER
展示中心景观

景观空间采用与建筑呼应的圆形中心环抱式结构，核心区域以"中国结"鼓穗作为视觉中心，配合高旱喷，气势宏伟、磅礴，强调向心凝聚力；鼓穗动感飘逸，与人行动线结合，形成明确的导向指引；盆景种植池采用"鼓状"造型，与建筑遥相呼应；广场边缘提炼中式"祥云"造型纹样；水景中间配置八幅安徽铁画，将著名的"合肥八景"浓缩其中，使万达文化与合肥地域文化实现了完美结合。

Landscape space uses echo circular center encircling structure consistent with the building and takes the "Chinese Knot" drum tassel as the visual center of the core area, presenting impressive, majestic, cohesive atmosphere coupled with high dry spray. The combination of dynamic and elegant drum tassel and pedestrian circulation forms clear orientation guide; the drum shaped bonsai planting trough echoes with the building afar; the plaza edge adopts Chinese style "Auspicious Cloud" pattern; waterscape is furnished with eight Anhui iron pictures epitomizing the noted "Eight Views of Hefei", achieving perfect fusion of Wanda culture and regional culture of Hefei.

INTERIOR OF EXHIBITION CENTER
展示中心内装

展示中心整体设计以"鼓"的元素贯穿建筑和室内。室内顶棚由一面巨大的鼓和许多小鼓造型组合而成，饰以龙纹发光图案，使建筑室内外形意相通。室内设计从墙身造型到地面图案，处处体现了徽派建筑的特色。脸谱、剪纸、竹简、印章等传统的中式元素，用现代的手法加以艺术处理，使其体现浓郁的"中国"韵味，达到了用现代的手法传达中国古典意蕴的设计意图。

The overall design of exhibition center employs the "Drum" element all the way through the building and its interior. The interior ceiling is shaped by a giant drum and a number of side drums with glowing dragon pattern, realizing interior-exterior consistency in shape and artistic concept. Anhui-style building features are found everywhere from the wall shape to the ground pattern. Subject to artistic treatment with contemporary technique, those traditional Chinese style elements like facial makeup, paper cutting, bamboo slip and seal deliver strong "Chinese" appeal, accomplishing the design idea of conveying the classical Chinese connotation with contemporary gimmick.

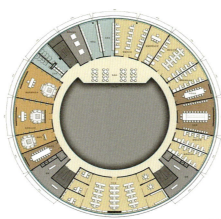

06

03 南昌万达城展示中心入口景观
04 南昌万达城展示中心景观
05 南昌万达城展示中心景观
06 南昌万达城展示中心平面图
07 南昌万达城展示中心大厅内装

07

2014 WANDA COMMERCIAL PLANNING
万达商业规划——销售类物业

04 展示中心

QINGDAO ORIENTAL CINEMA EXHIBITION CENTER
青岛东方影都展示中心

开放时间 2013 / 11 / 16	**OPENED ON** NOVEMBER 16 / 2013
项目位置 山东 / 青岛	**LOCATION** QINGDAO / SHANDONG PROVINCE
占地面积 1.50 公顷	**LAND AREA** 1.50 HECTARES
建筑面积 0.41 万平方米	**FLOOR AREA** 4,100m²

PART B CULTURAL TOURISM PROJECT
文旅项目

01 青岛东方影都展示中心鸟瞰图
02 青岛东方影都展示中心夜景
03 青岛东方影都展示中心总平面图

BUILDING OF EXHIBITION CENTER
展示中心建筑

展示中心是青岛东方影都项目首个标志性建筑。建筑借鉴了海洋"活化石"鹦鹉螺的造型，以海洋生物形态作为设计概念，意在展示青岛这座著名海滨城市的鲜明地域特点和历史文化底蕴。鹦鹉螺形体优美的渐开线"映射"青岛美丽、浪漫、现代的城市特征，并且实现了建筑形态与使用功能完美结合，形成具有万达特色的滨海建筑形象。

As the first landmark building of Qingdao Oriental Cinema Project, the Exhibition Center building, drawing on the modeling of Nautilus, the Marine "living fossil", takes the marine life form as its design concept, attempting to demonstrate the distinct regional features and historical & cultural foundation of Qingdao, a famous coastal city. The gracefully shaped involute of nautilus "maps" the city characters of Qingdao, being beautiful, romantic and modern, and perfectly combines the architectural form and use function, taking on a seafront building image with Wanda features.

LANDSCAPE OF EXHIBITION CENTER
展示中心景观

设计以"海螺"轮廓线规划景观广场，将青岛的阳光、沙滩、红礁石及海滩上的晨雾等地域形象融入其中，与建筑外形呈镜像关系，相互呼应。栈道、沙滩、水池与广场为流畅的"S"形设计，从"鹦鹉螺"螺口延伸向外，再现海滩、浪花、码头等意向，地面铺装呈螺旋状向外层层延伸。展示中心如同浸泡在海水中的鹦鹉螺一般，结合《侏罗纪公园》、《变形金刚》、《木马屠城》、《非诚勿扰》等影视主题雕塑，营造出东方好莱坞迷人的当代影都风情。

Planned as per "conch" contour line and integrated with sunshine, beach, red reef and the morning mist on the beach in Qingdao, the landscape plaza boasts of mirror-image relation and harmonious coexistence with architectural appearance. The smooth S shaped gallery road, beach, pool and plaza square extend outward from the "Nautilus" mouth to reproduce the images of beaches, waves, wharf, and floor pavement is spiraling outward layer upon layer, making the exhibition center looked as if a nautilus soaked in the water. Combining with film and television theme sculptures such as Jurassic Park, Transformers, Troy and if You Are the One, appealing contemporary movie metropolis style of the oriental Hollywood is thus felt.

04

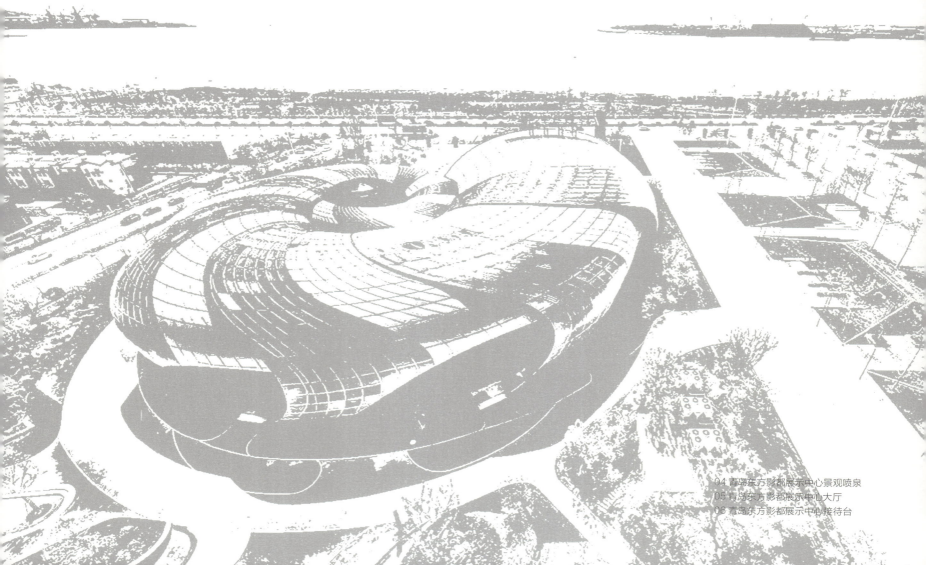

04 青岛东方影都展示中心景观喷泉
05 青岛东方影都展示中心大厅
06 青岛东方影都展示中心接待台

INTERIOR OF EXHIBITION CENTER
展示中心内装

室内设计以"海—岛—藻—水—陆—树"渐次推进的设计概念，打造出一个美轮美奂、奇异而沉静的空间；墙身、挂件、陈设融合经典影片《变形金刚》、《复仇者联盟》中的角色形象，紧扣东方影都的文化主题，通过高科技声、光、电、影的设计，丰富客户体验，提升展示效果。

Interior design, following the design concept of gradually advancing "sea-island-algae-water-land-tree", builds an incredible, extraordinary and tranquil space. Adding the role images from classic movies of Transformers and The Avengers, the wall body, pendants, furnishings sick closely to the cultural theme of the Oriental Cinema, and high-tech sound, light, electricity and image design is available, both contributing to enriching customers' experience and improving the exhibition effect.

2014 WANDA COMMERCIAL PLANNING
万达商业规划——销售类物业

展示中心
05

WUXI WANDA CITY EXHIBITION CENTER
无锡万达城展示中心

开放时间 2014 / 03 / 25	**OPENED ON** MARCH 25 / 2014
项目位置 江苏 / 无锡	**LOCATION** WUXI / JIANGSU PROVINCE
占地面积 0.58 公顷	**LAND AREA** 0.58 HECTARES
建筑面积 0.43 万平方米	**FLOOR AREA** 4,300m²

BUILDING OF EXHIBITION CENTER
展示中心建筑

展示中心以"壶天盛景"为设计主题，运用先进的BIM技术，将无锡历史悠久的地方民俗（陶艺文化、茶文化）融入建筑设计。选用棕红色铝板与金属穿孔板建筑表皮，与新风系统有机结合。为保证中庭体验、减轻结构自重、控制成本，运用传统江南园林设计"框景"手法将建筑物中心部分非功能区去除。场地设计巧妙利用地势高差，将建筑首层抬高2.5米，增大首层面积，满足展示功能的需求，突出展示中心庄重大气的入口形象；同时减少了地下室的土方开挖量和工程降水量，避开了无锡特有的淤泥层，取得了节约成本、规避风险、加快工程进度的效果。

Following the design theme of "Grand Pottery View", the Exhibition Center employs advanced BIM technologies to blend the long-standing local folk of Wuxi (ceramic culture and tea culture) into architectural design. The building surface chooses palm red aluminum plate and metal perforated plate and realizes organic combination with fresh air system. To ensure the atrium experience, reduce self-weight and control the cost, the design adopts "enframed scenery" technique available in traditional south China garden design to remove the nonfunctional area in building center. The site design, with ingenious utilization of terrain elevation difference, raises the building ground floor for 2.5 meters to increase of its area and meet the needs of exhibition function, highlighting the grand and dignified entrance image of exhibition center and at the same time, reducing the basement earthwork excavated volume and engineering precipitation by avoiding the silt layer specific to Wuxi. In this manner, the design helps to save cost, mitigate risk and speed up the project progress.

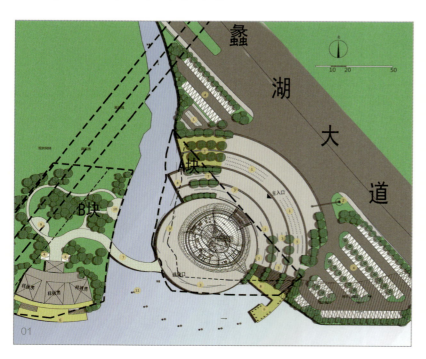

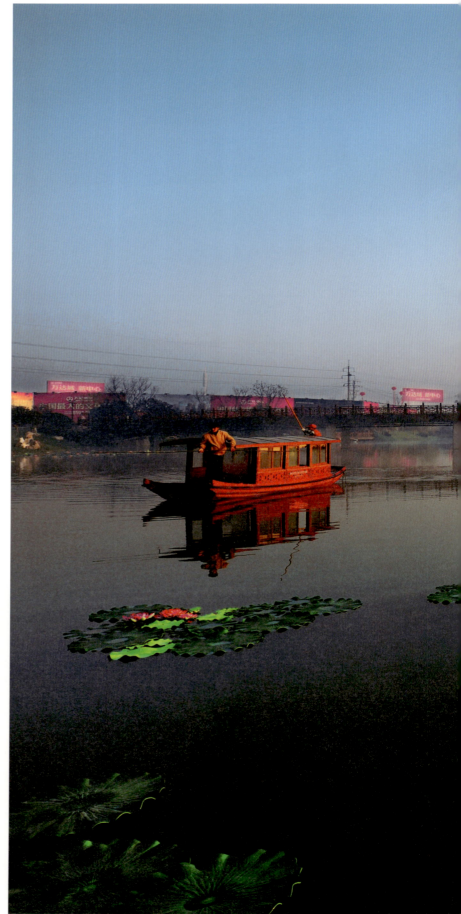

PART B | CULTURAL TOURISM PROJECT
文旅项目

01 无锡万达城展示中心总平面图
02 无锡万达城展示中心建筑远景

2014 WANDA COMMERCIAL PLANNING
万达商业规划——销售类物业

LANDSCAPE OF EXHIBITION CENTER
展示中心景观

以江南园林为主题，充分利用周围的景观资源，借地、借山、借水、借景，"师法自然，融于自然"。建造面向雪浪山国家公园方向、高于室外地面1.5米的景观平台，使雪浪山国家公园成为展示中心的城市对景。对河道进行改造、清淤，种植莲花、芦苇等水生植物，改善局部生态环境，将河对岸的景观改造为以原生植被为主的生态湿地，形成对岸岸线景观作为远景，河面莲花、芦苇作为近景的多层次景观。

Focusing on the theme of South China garden, the landscape makes full use of the surrounding landscape resources, such as land, mountain, water and scenery, to be inspired by and assimilated into nature. Through building a landscape platform that is 1.5 meters higher than the outdoor ground and faces Xuelang Mountain Ecological and Scenic Garden, the garden presents the urban opposite scenery of the exhibition center; through transformation, dredging of river channel and planting of lotus, reeds and other aquatic plants for improving the local ecological environment, the landscape across the river is transformed into a ecological wetland with overwhelming native vegetation. Consequently, a multi-level landscape is available with the opposite bank landscape being a distant view and the lotus and reed on the river as a nearby view.

PART B CULTURAL TOURISM PROJECT
文旅项目

INTERIOR OF EXHIBITION CENTER
展示中心内装

以江南水乡文化为主题——彩色穹顶中心为"太湖明珠",外圈分别为"鱼米之乡"、"状元及第"等抽象意向。彩色玻璃穹顶烘托了恢宏大气的室内空间,与巨大的中心沙盘互相衬托、相得益彰。展示中心内墙面采用了多孔吸声板,表面饰以凹凸纹理墙布,通过计算机模拟以及现场实测,声音传输指数得到改善,大空间实现了良好的声学效果。

Taking the theme of South Yangtze River water town culture, the interior depicts abstract images, including "Pearl of Taihu Lake" in center and "A Land of Fish and Rice" and "Number One Scholar" outside of colored dome. The stained glass dome is served as a foil of the grand interior space, and achieves harmonious coexistence with the central building model. With porous sound-absorbing board applied in inner wall and surface decorated with concave and convex textured wall cloth, and benefiting from the computer simulation and field measurement, the exhibition center attains improved sound transmission index and desirable acoustic effects in the large space.

05 无锡万达城展示中心大厅
06 无锡万达城展示中心沙盘
07 无锡万达城展示中心沙盘俯视

2014 WANDA COMMERCIAL PLANNING
万达商业规划——销售类物业

06 展示中心

GUANGZHOU WANDA CITY EXHIBITION CENTER
广州万达城展示中心

开放时间 2014 / 12 / 06	**OPENED ON** DECEMBER 6 / 2014
项目位置 广东 / 广州	**LOCATION** GUANGZHOU / GUANGDONG PROVINCE
占地面积 1.00 公顷	**LAND AREA** 1.00 HECTARES
建筑面积 0.60 万平方米	**FLOOR AREA** 6,000m²

01 广州万达城展示中心总平面图
02 广州万达城展示中心建筑设计意向图
03 广州万达城展示中心夜景

BUILDING OF EXHIBITION CENTER
展示中心建筑

广州万达城展示中心以"木棉花"为主题，十二片形态优美的三维曲面重复、旋转，形成一支含苞欲放的美丽花朵。"花"的主题为展示中心找到了一个恰如其分的、具象的地域性符号，更能唤起广州市民的文化认同。"12"代表12个月份，"红色"代表广州的热情与朝气，含苞待放的形态象征生机勃勃的张力。

Through repetition and rotation of twelve gracefully-shaped 3D curved surfaces, a beautiful budding flower and "Kapok" theme are presented. Taking "Kapok" as the theme provides the exhibition center with a fit and specific regional symbol, and is more likely to arouse the cultural empathy of Guangzhou citizens. In the design, the "twelve" surfaces stand for 12 months, the "red" color symbolizes the enthusiasm and vigor of Guangzhou City, and the budding image delivers the dynamic tension.

01

PART B CULTURAL TOURISM PROJECT
文旅项目

02

03

04

04 广州万达城展示中心景观
05 广州万达城展示中心景观雕塑
06 广州万达城展示中心景观绿化
07 广州万达城展示中心水景

LANDSCAPE OF EXHIBITION CENTER
展示中心景观

"挥毫泼墨蕴盛世，撞水撞粉笔生花"——景观设计概念源于岭南画派，通过运用"撞水、撞粉"技法，结合建筑造型的肌理，以建筑为"主花"，围绕建筑簇拥着不同功能的"小花"，晕染出层层渐变的景观空间，展现出韵味十足、生生不息的"木棉花开图"。亲水木栈道犹如建筑衍生出来的花瓣，漂浮在水面，延伸至样板间，客户体验过程中可尽享宽屏湖景视野，实现"让建筑吸引人群，让景观留住人群"的设计初衷。

Sourcing its idea from Lingnan Painting school that "Using Painting To Imply the Flourishing Age and Adding Water or Powder to Still-drying Pigment to Realize Outstanding Painting Technique", the landscape design, through employing the technique concerned and combining the texture of architectural modeling, takes the building as the "main flower" surrounded by "flowerlets" with diverse functions. Thus, a gradient landscape space by layer is created, depicting the flourishing "Blooming Kapok Picture" with lasting appeal. In addition, the hydrophile wooden viaduct looks as if the petals of the building floating on water and extending toward the mockup room, where customers can enjoy the spacious lake view. Thus, the original design intention of "Buildings to Attract and Landscape to Keep Customers" is realized.

PART B CULTURAL TOURISM PROJECT
文旅项目

05

06

07

PART B　CULTURAL TOURISM PROJECT
文旅项目

2014 WANDA COMMERCIAL PLANNING
万达商业规划——销售类物业

样板间 **01**

PROTOTYPE ROOM OF HARBIN WANDA CITY
哈尔滨万达城样板间

开放时间 2014 / 07 / 19	OPENED ON JULY 19 / 2014
项目位置 黑龙江 / 哈尔滨	LOCATION HARBIN / HEILONGJIANG PROVINCE
建筑面积 130 平方米	FLOOR AREA 130m²

FRENCH NEOCLASSICAL STYLE PROTOTYPE ROOM
法式新古典风格样板间

内装以新古典主义法式风格为基调，墙壁浅咖色红影木造型饰面，为空间确立了温暖的主环境色。主背景墙选择手绘花鸟壁纸活跃了空间氛围，也增加了材质层次。简单造型的吊顶，既迎合了新古典的空间气质，也弱化了传统古典风格繁复叠加的压抑感。

Keeping to the keynote of French neoclassical style, the interior design applies light coffee color anegre-shaped finish to the wall for establishing a warm primary environmental color for the space, selects hand-painted flowers and birds wallpaper for main background wall to present active space atmosphere and increased material levels, and employs simply shaped ceiling to cater for neoclassical space temperament and weaken the sense of repression brought about by complicated superposition found in traditional classical style.

01 哈尔滨万达城样板间书房
02 哈尔滨万达城样板间客厅
03 哈尔滨万达城样板间卧室

2014 WANDA COMMERCIAL PLANNING
万达商业规划——销售类物业

样板间 **02**

PROTOTYPE ROOM OF WUHAN CENTRAL CULTURE DISTRICT
武汉中央文化区样板间

开放时间 2014/05/08	OPENED ON MAY 8 / 2014
项目位置 湖北/武汉	LOCATION WUHAN / HUBEI PROVINCE
建筑面积 1296 平方米	FLOOR AREA 1,296m²

01 武汉中央文化区样板间-法式新古典风格样板间客厅
02 武汉中央文化区样板间-法式新古典风格样板间门厅
03 武汉中央文化区样板间-法式新古典风格样板间卧室

01

FRENCH NEOCLASSICAL STYLE PROTOTYPE ROOM
法式新古典风格样板间

温暖、淡雅的主色调，辅以法式雕刻的典雅顶棚、"点睛之笔"的金色装饰，营造出法式浪漫空间。天然大理石、精雕细琢的线脚、华丽的壁纸、大型灯池，处处体现精致与豪华，搭配法式新古典家具和饰品，再现十八世纪法国宫廷生活空间。

The warm and subdued primary color, supplemented with the French carved elegant ceiling and golden decoration with punch line effect builds a French romantic space. Delicacy and luxury are found everywhere from natural marble to crafted molding and from ornate wallpaper to large lamp pool. Going along with French neoclassical furniture and accessories, the 18th Century French court life space is represented.

GRADE A OFFICE BUILDING PROTOTYPE ROOM
甲级写字楼

空间定位为高端金融企业总部办公。功能布局采用直线分割手法，动静分区，沉稳大气。高档名贵的进口棕色石材、经"钢琴漆"表面工艺处理的天然红木、纯铜锻造的屏风，在灯光映射下流光溢彩，色调沉稳高贵，彰显高端企业雄厚实力。

As it is positioned as the high-end financial headquarter office space, the functional layout adopts line segmentation technique to separate the living and the privacy areas and present composed and grand atmosphere. Moreover, the high-grade imported brown stone, natural rosewood treated with piano lacquer technique turn to be colorful, calm and noble under the light, mirroring the abundant strength of high-end enterprise.

04 武汉中央文化区样板间-甲级写字楼总裁办公室
05 武汉中央文化区样板间-甲级写字楼前台
06 武汉中央文化区样板间-甲级写字楼会议室

2014 WANDA COMMERCIAL PLANNING
万达商业规划——销售类物业

样板间 **03**

PROTOTYPE ROOM OF NANCHANG WANDA CITY
南昌万达城样板间

开放时间 2014 / 05 / 17	**OPENED ON** MAY 17 / 2014
项目位置 江西 / 南昌	**LOCATION** NANCHANG / JIANGXI PROVINCE
建筑面积 772 平方米	**FLOOR AREA** 772m²

GEORGIA STYLE PROTOTYPE ROOM
乔治亚风格样板间

十八世纪至十九世纪，乔治亚风格在欧洲（特别是英国）流行，对美国别墅风格也产生明显影响。本套样板间设计严格对称、比例均衡，选用了传统的墙裙、墙面、檐壁的"三段式"墙面处理方式，具有典型的乔治亚风格。在此基础上，保留了欧式古典的色泽和质感，强调简洁、明晰的线条和优雅、有度的装饰，使空间变得更加轻松、舒适。

During the 18th and 19th centuries, the Georgia Style was prevailing in Europe (especially UK) and has a great presence in villa style of the United States. In rigorous symmetry and balanced proportion, the design of this prototype room applies the traditional "three-segment" (including dado, metope and frieze) wall treatment approach, endowing it with typical Georgia style. Based on this, European classic color and texture are maintained by highlighting concise & clear lines and elegant & modest adornment, presenting a more relaxed and comfortable space.

01 南昌万达城样板间-乔治亚风格样板间卧室
02 南昌万达城样板间-乔治亚风格样板间客厅

2014 WANDA COMMERCIAL PLANNING
万达商业规划——销售类物业

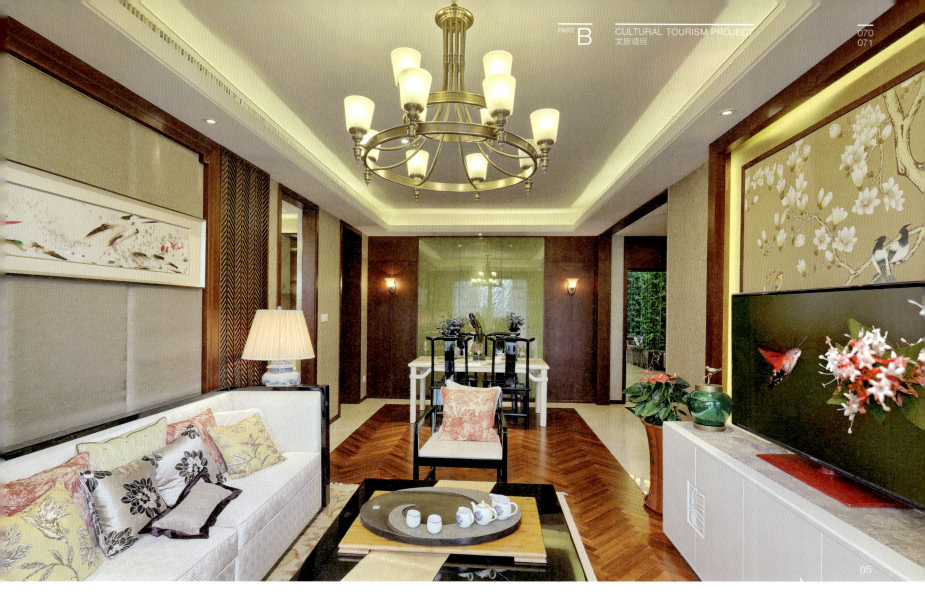

03 南昌万达城样板间－乔治亚风格样板间客厅
04 南昌万达城样板间－乔治亚风格样板间客厅
05 南昌万达城样板间－新中式风格样板间客厅
06 南昌万达城样板间－新中式风格样板间卧室

NEW CHINESE STYLE PROTOTYPE ROOM
新中式风格样板间

设计采用现代手法演绎空间，将传统元素加入其中，使之更能符合现代人的审美及舒适要求。黑檀木、米白色大理石和高光漆的古典中式家具，进一步强化了设计主题，精致的手绘壁纸软包与木线条结合提升了空间品质。

While embracing traditional elements, the design chooses to interpret the space with modern technique, making it more close to the modern people's aesthetic and comfortable demands. Ebony, rice white marble and high gloss classical Chinese style furniture further strengthens the design theme, and the combination of delicate hand-painted wallpaper and timber moldings improve the quality of space.

2014 WANDA COMMERCIAL PLANNING
万达商业规划——销售类物业

样板间 04

PROTOTYPE ROOM OF WUXI WANDA CITY
无锡万达城样板间

开放时间 2014 / 09 / 26　　**OPENED ON** SEPTEMBER 26 / 2014
项目位置 江苏 / 无锡　　　**LOCATION** WUXI / JIANGSU PROVINCE
建筑面积 636 平方米　　　**FLOOR AREA** 636m²

01 无锡万达城样板间－法式风格样板间客厅
02 无锡万达城样板间－法式风格样板间中庭
03 无锡万达城样板间－法式风格样板间卧室

02

03

FRENCH STYLE PROTOTYPE ROOM
法式风格样板间

整体风格沉稳、气派，空间宽敞明亮、富丽堂皇；晶莹剔透的米黄大理石、典雅高贵的水晶吊灯及古典欧式雕花等经典的法式元素塑造出欧式城堡生活空间；大面积菱形拼花地面彰显气派空间，与菱形角线上下呼应，格调统一；精心选择的家具饰品，流露出厚重的历史沉淀感。

The whole room enjoys a composed & gorgeous style and capacious & splendid space. Classic French elements of glittering and translucent cream-colored marble, elegant and noble crystal droplight and classical European carving help to create a European castle life space; large-sized diamond parquet floor highlighting gorgeous space echoes with upper diamond corner moldings to form a unified style. Carefully selected furniture accessories reveal the profound historical accumulation.

04 无锡万达城样板间-简欧风格样板间客厅
05 无锡万达城样板间-简欧风格样板间客厅一角
06 无锡万达城样板间-简欧风格样板间卧室
07 无锡万达城样板间-新中式风格样板间餐厅
08 无锡万达城样板间-新中式风格样板间客厅

SIMPLIFIED EUROPEAN STYLE PROTOTYPE ROOM
简欧风格样板间

温暖的红色调为主，细致、简约的欧式风格贯穿整个空间。白色木作、米黄石材、繁花地毯，营造出经典华丽的空间。细腻的装饰线条、精挑细选的欧式古典家具及隔断造型，在提升质感的同时也表达了对古典文化的尊重与追求。

NEW CHINESE STYLE PROTOTYPE ROOM
新中式风格样板间

取自天然的设计素材最得传统中式神韵，赋予优雅的东方气质。中国风壁纸与深色木作相得益彰；干净简洁的白色石材地面拼花成功地烘托出墙面装饰、家具及饰品的质感；流畅协调的格局实现了视觉净化，凸显平面布局的通透和开放。这些风格把东方意象融入现代生活中，传达了中国式的优雅。

The design chooses natural elements that can best exhibit the traditional Chinese charm and elegant Oriental style. To be specific, Chinese style wallpaper and dark carpentry complement each other; clean and concise white stone floor parquet successfully foil the quality of wall adornment, furniture and accessories; smooth and coordinated layout realizes visual purification and highlights a transparent and open plane layout. These styles, through assimilating the Oriental images into modern life, convey the Chinese style elegance.

2014 WANDA COMMERCIAL PLANNING
万达商业规划——销售类物业

样板间 **05**

PROTOTYPE ROOM OF QINGDAO ORIENTAL CINEMA
青岛东方影都样板间

开放时间 2014 / 06 / 28	**OPENED ON** JUNE 28 / 2014
项目位置 山东 / 青岛	**LOCATION** QINGDAO / SHANDONG PROVINCE
建筑面积 480 平方米	**FLOOR AREA** 480m²

STAR OCEAN HALL
星海大厅

大厅以海洋为主题。顶棚造型模拟海洋潮汐,点缀海星及鱼形水晶吊灯;闪闪发光的贝壳喷泉、色彩艳丽的海浪波纹铺地,处处洋溢着浓郁的海洋气息。休息区再现了电影《哈利·波特》之"魔法商店"和"美国小镇"中的场景,紧扣东方影都项目定位。

Following the theme of ocean, the hall design adopts ocean tide-shaped ceiling embellished with starfish and fish shaped crystal chandeliers, sparkling shell fountains and colorful ocean waves floor pavement, making full-bodied marine atmosphere felt everywhere. As for the rest area, to closely conforming to the project positioning, the scenes of "Magic Shop" and "American Town" in the film "Harry Potter" are reproduced.

PART B | CULTURAL TOURISM PROJECT
文旅项目

01 青岛东方影都样板间-星海大厅
02 青岛东方影都样板间-星海大厅

MEDITERRANEAN STYLE PROTOTYPE ROOM
地中海风格样板间

设计灵感来源于影都"戛纳"。以清新的蓝绿色调为主，配以现代欧式家具和舒适柔软的布艺，使整个空间充满了夏日味道，仿佛置身戛纳海边悠闲地度假。空间功能划分明确，多种材质和材料融合穿插，丰富了空间感受。使用金色、黄色、暗红色等主色调的软装配饰，糅合少量白色，使整体空间明亮、优雅。

Inspired by "Cannes", the movie metropolis, the design gives priority to blue-green color and arranges modern European furniture, comfortable and soft fabrics to make the whole space full of summer flavor, offering the experience of enjoying holiday by Cannes seaside. Clear functional division fused and intertwined with a variety of materials and textures enrich the space experience. Fitted with gold, yellow, dark red and fragmented white accessories, the whole space becomes bright and elegant.

CLASSICAL EUROPEAN STYLE PROTOTYPE ROOM
古典欧式样板间

设计灵感来源于水城威尼斯。在室内家具和饰品的选用上，以优雅舒适的暖色调为主——精致的描金线条、古典欧式家具，搭配丝绸般光滑触感的布艺——使空间层次丰富，视觉感受精致而华丽。在色彩搭配上，注重统一协调，运用典雅的深蓝色渲染出威尼斯水城特色。

Sourcing its idea from Venice, the city of water, the design mainly follows the elegant and comfortable warm colors in selecting indoor furniture and accessories. The adoption of delicate gold-outlined lines, classical European-style furniture and silky smooth fabrics helps to present a space with rich gradation and a delicate and gorgeous visual perception. With regard to color collocation emphasizing harmony, elegant dark blue color is employed to render the features of the water city.

03 青岛东方影都样板间-地中海风格样板间客厅
04 青岛东方影都样板间-地中海风格样板间卧室
05 青岛东方影都样板间-古典欧式风格样板间卧室
06 青岛东方影都样板间-古典欧式风格样板间客厅

2014 WANDA COMMERCIAL PLANNING
万达商业规划——销售类物业

样板间 06

PROTOTYPE ROOMS OF GUANGZHOU WANDA CITY
广州万达城样板间

开放时间 2014 / 12 / 06	**OPENED ON** DECEMBER 6 / 2014
项目位置 广东 / 广州	**LOCATION** GUANGZHOU / GUANGDONG PROVINCE
建筑面积 318 平方米	**FLOOR AREA** 318m²

02

01

APARTMENT PROTOTYPE ROOM
公寓样板间

空间的比例对称、均衡，明朗的线条简洁、流畅；规矩的序列传达出中式古韵的严谨。原木家具、饰品散发自然的气息，彰显了环保的理念。色彩搭配以纯色为基调，无论是墙面、地面还是装饰摆件，都显示出单一色调所带来的简约、流畅之感。

The spatial proportion is symmetric and well-balanced, and the clear lines are succinct and smooth; the orderly layout reflects the preciseness of Chinese-style ancient rhyme. Furniture and decorations made of raw woods emit a breath of nature and manifest the concept of environment protection. As pure colors are taken as the keynote of color matching, the walls, the floor and the decorations together reflect a sense of simplicity and smoothness brought by a unique tone.

01 广州万达城样板间－公寓样板间卧室
02 广州万达城样板间－公寓样板间电梯间
03 广州万达城样板间－公寓样板间卧室

LEATHER COMPANY PROTOTYPE ROOM
皮革公司样板间

根据广州当地商业特征定位，使用二合一空间打造时尚高端的皮具公司样板间，为目标客户提供置业参考。强调弧形的入口区域，体现商务氛围。墙面用皮革与精钢及发光片组合而成LOGO墙，体现出皮具产品展示空间的亲人性及高贵性。展示区墙面的产品展示与手工定制台的立体型展示，可以让客户身临其境感受到纯手工制品的高端品质。

In accordance with the local commercial characteristics of Guangzhou, a fashionable and high-end prototype room for leather company is designed with a two-in-one space, providing a purchasing reference to target clients. The cambered entrance area reflects a business atmosphere. A LOGO wall made up of leather, stainless steel and light panels manifests the intimacy and nobleness of the exhibition space of leather products. Through the product display on the walls of the exhibition area and the three-dimensional display on the manual custom stand, customers can feel the high quality of purely hand-made products personally.

04

04 广州万达城样板间-皮革公司样板间户型图
05 广州万达城样板间-皮革公司样板间
06 广州万达城样板间-首饰公司样板间
07 广州万达城样板间-首饰公司样板间楼梯

JEWELRY COMPANY PROTOTYPE ROOM
首饰公司样板间

通过结构设计的精确核算，创造出超大悬挑空间，极大提升空间价值。挑高空间圆弧形楼梯在展现出空间柔美感的同时，与珠宝的产品属性完美结合。墙面面层采用浅咖色高档铝板与高光烤漆板相结合，配合专业的空间照明设计满足珠宝展示对于施工工艺、灯光照明的严苛要求，曲面及柱型展台既满足产品展示效果，又能够满足散热的功能要求。

Based on a precise calculation of structural design, a long cantilevered space is created, which largely elevates the value of the space. The circular stairs in the cantilevered space not only reveals the grace of the space but also perfectly matches the attributes of jewelries products. The wall surface in combination of light coffee quality aluminum plate and specular high-gloss lacquering board, together with a professional space lighting design satisfies the jewelry exhibition's harsh requirements on construction techniques and lighting. Meanwhile, the cambered and column-type exhibition stand accommodates not only the exhibition effects need of products but also the functional requirements of thermal dissipation.

2014 WANDA COMMERCIAL PLANNING
万达商业规划——销售类物业

样板间
07

PROTOTYPE ROOMS OF HEFEI WANDA CITY
合肥万达城样板间

开放时间 2014 / 12 / 28　　**OPENED ON** DECEMBER 28 / 2014
项目位置 安徽 / 合肥　　**LOCATION** HEFEI / ANHUI PROVINCE
建筑面积 380 平方米　　**FLOOR AREA** 380m²

01 合肥万达城样板间-法式风格样板间客厅
02 合肥万达城样板间-法式风格样板间卧室
03 合肥万达城样板间-法式风格样板间餐厅
04 合肥万达城样板间-法式风格样板间书房

FRENCH-STYLE PROTOTYPE ROOM
法式风格样板间

平面布局突出轴线对称；米白色调亮光漆造型木挂板，营造了淡雅的家居风格，也增强了视觉的宽阔感；冷色系碎花窗帘、布艺及墙体壁纸，相互呼应；家具造型简洁大气，细节处理注重雕花和线条的制作工艺，局部描金处理，凸显法式浪漫奢华气息，充分彰显主人之高贵身份与地位。

The plan layout highlights axis symmetry; the creamy-white brilliant varnish wooden hanging plate renders a simple and elegant decorating style and enhances the visual impression of broadness as well; the cool color floral curtain and fabrics echo with the wall paper; the modeling of the furniture is concise and grand and the details treatment is found in the craftsmanship of carvings and lines, with partial outlines in gold, which as a whole reflect a romantic and luxurious French-style charm, fully manifesting the gentility and noble status of the owner.

AMERICAN-STYLE PROTOTYPE ROOM
美式风格样板间

白色的木饰造型和米色仿古砖铺地结合，营造了自在、休闲的空间氛围。家具、饰品强调优雅的雕刻和精致的细节处理，在保留古典家具的体量和质感的同时，又注意适应现代生活空间，力求营造自在与随意的美式风格。

The combination of white wooden modeling and beige archaized brick pavement creates a cozy and casual space atmosphere. The furniture and decorations highlight exquisite carvings and delicate detail processing for making the room an appropriate modern living space while preserving the volume and quality of classical furniture, striving to present an easy and casual American style.

05 合肥万达城样板间－美式风格样板间卧室
06 合肥万达城样板间－美式风格样板间客厅
07 合肥万达城样板间－田园风格样板间客厅
08 合肥万达城样板间－田园风格样板间卧室
09 合肥万达城样板间－田园风格样板间阳台一角

PASTORAL STYLE PROTOTYPE ROOM
田园风格样板间

开放的空间结构、随处可见的花卉和绿色植物、雕刻精细的家具……回归田园的清新气息扑面而来。装饰着太阳花的草帽、窗前微微晃动的摇椅，在任何一个角落，都能体会到主人悠然自得的生活和阳光般明媚的心情。

The open space structure, the prevailing flowers and green plants and the delicately-carved furniture all render a fresh breath that brings people back to pastoral life. From the straw hat decorated with sunflower to the slightly swaying rocking chair in front of the window, the owner's carefree and content life and his bright and graceful mood are felt at any corner of the room.

2014 WANDA COMMERCIAL PLANNING
万达商业规划——销售类物业

实景示范区 01

DEMONSTRATION AREA OF HARBIN WANDA CITY
哈尔滨万达城实景示范区

开放时间 2014 / 07 / 16	OPENED ON JULY 16 / 2014
项目位置 黑龙江 / 哈尔滨	LOCATION HARBIN / HEILONGJIANG PROVINCE
占地面积 0.98 公顷	LAND AREA 0.98 HECTARE

PLANNING OF DEMONSTRATION AREA
示范区建筑规划

实景示范区整体布局分为入口形象区、中心景观区、休闲活动区及样板房景观区四个部分，包含百米高层住宅和150米超高层住宅。建筑采用简欧风格立面，通过对比例、尺度、细部的精细化设计，较好地体现了欧式建筑的品质感。整体色调明快，能够削弱高层建筑巨大的体量感；局部穿插深咖、浅灰色涂料，起到活泼立面的效果；基座部分采用深咖涂料，视觉上具有稳定之感。

The demonstration area is generally laid out with entrance image area, central landscape area, leisure activity area and prototype room landscape area, encompassing a 100-meter high-rise residence and a 150-meter ultra high-rise residence. The buildings adopt simple European-style facade, and through an exquisite design of proportion, scale and details, the quality sense of European-style architecture is well achieved. The overall bright and lively colors of the buildings successfully mitigate the huge volume sense brought by the high-rise buildings, the partially interpersed dark coffee and light gray paintings enliven the facade, and the dark coffee painting applied on the foundation part renders a visual impression of stability.

01

02

PART B | CULTURAL TOURISM PROJECT
文旅项目

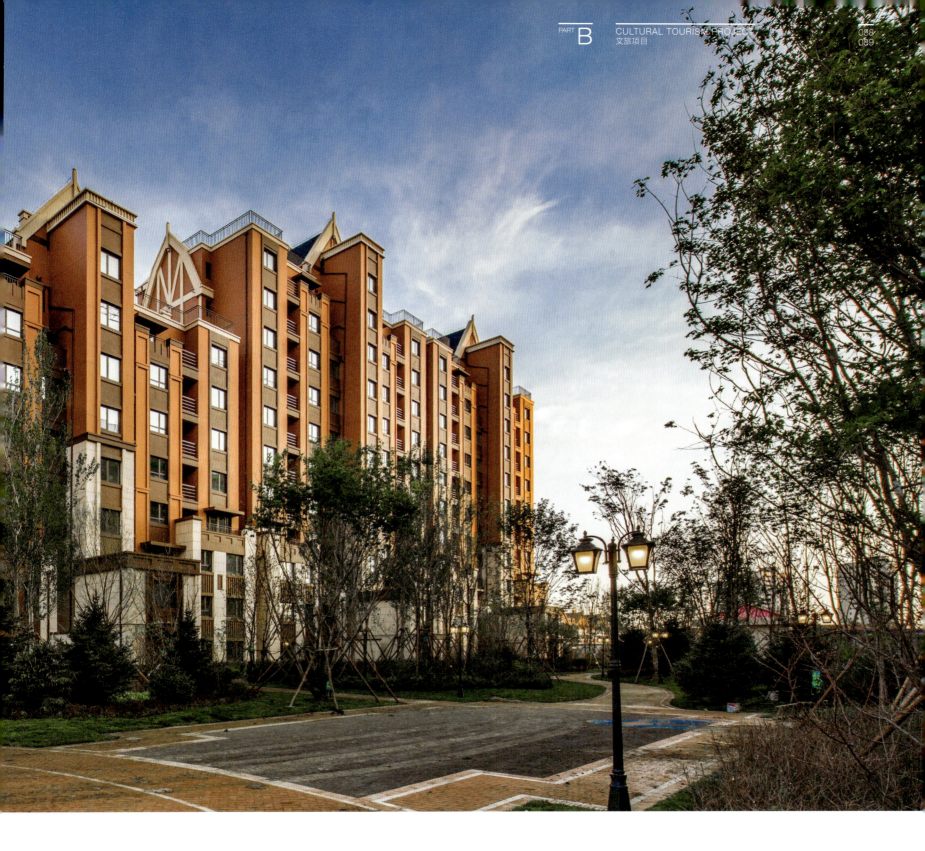

01 哈尔滨万达城实景示范区建筑外立面
02 哈尔滨万达城实景示范区总平面图
03 哈尔滨万达城实景示范区建筑外立面

LANDSCAPE OF DEMONSTRATION AREA
示范区景观

设计风格沿袭欧式建筑风格，打造公园式景观体验区。入口形象区采用序列、对称花钵，结合水景和特色树池，展现出入口景观区的仪式感和庄重感。拾阶而上的中心水景广场、舒适的坡地白桦林、盛开的各色鲜花、叮咚奏鸣的潺潺流水，为人们提供休憩、观光的场所。曲折流动的林荫通道、自然起伏的坡地景观、活动休闲景观亭，营造出功能齐全、氛围轻松的休闲活动区。

The garden-like landscape experience area is designed by following the style of European architecture. In the entrance image area, arrays of symmetric flowerpots, combined with waterscape and feature tree pools, reflect the ritual sense and dignity of the entrance landscape area. In the area for relaxation and sight-seeing, there is a stepped central waterscape plaza, a cozy birch forest in a sloping land, blooming flowers of various colors and a babbling cascading waterfall. The circuitous tree-lined paths, the landscape of naturally undulating sloping field and the landscape pavilion for activities and leisure as a whole create a well-functioned and relaxed leisure activity area.

04 哈尔滨万达城实景示范区水景
05 哈尔滨万达城实景示范区景观绿化
06 哈尔滨万达城实景示范区景观亭

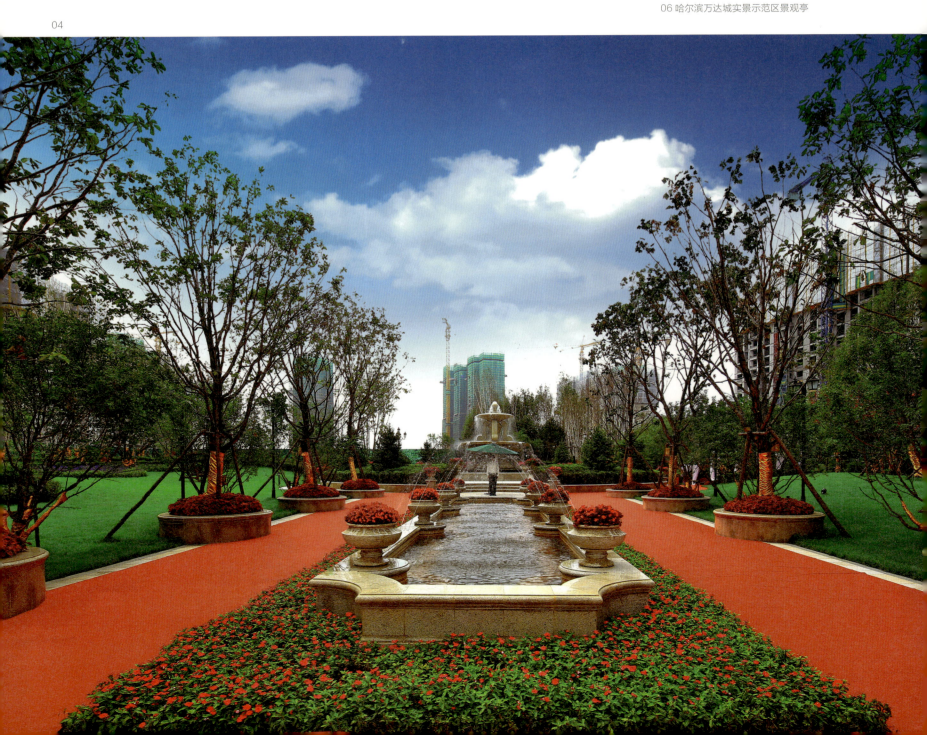

PART B CULTURAL TOURISM PROJECT
文旅项目

2014 WANDA COMMERCIAL PLANNING
万达商业规划——销售类物业

实景示范区
02

DEMONSTRATION AREA OF NANCHANG WANDA CITY
南昌万达城实景示范区

开放时间 2014 / 05 / 18　　OPENED ON MAY 18 / 2014
项目位置 江西 / 南昌　　　LOCATION NANCHANG / JIANGXI PROVINCE
占地面积 0.97 公顷　　　　LAND AREA 0.97 HECTARE

PLANNING OF DEMONSTRATION AREA
示范区建筑规划

南昌万达城实景示范区整体布局分为入口展示区、中心景观区、休闲活动区及样板房展示区四个部分，是万达城B区双拼及联排别墅样板展示区。本项目采用新唐风古典立面，通过对比例、尺度、细部的精细化设计，较好地体现了中式建筑的品质感。整体色调古典端重，外墙采用干挂石材，辅以深褐色真石漆等材质，视觉上不仅具有稳定之感，且有典雅之韵味。

The landscape demonstration area of Nanchang Wanda City is generally laid out with entrance image area, central landscape area, leisure activity area and prototype room landscape area, serving as a prototype demonstration area for the duplexes and united villas in Zone B of the Wanda Cultural Tourism City. This project adopts the new Tang style classical facade, and through an exquisite design of proportion, scale and details, the quality sense of Chinese-style architecture is well achieved. The overall color of the building takes on a classical and dignified atmosphere. The dry-hang stone external wall, supplemented with dark brown stone-like coatings, renders a visual sense of stability as well as a lingering charm of elegance.

01

PART B | CULTURAL TOURISM PROJECT
文旅项目

01 南昌万达城实景示范区总平面图
02 南昌万达城实景示范区建筑外立面
03 南昌万达城实景示范区建筑外立面

LANDSCAPE OF DEMONSTRATION AREA
示范区景观

设计采用中式景观风格，利用溪流水系、观景木平台、假山、植栽等元素形成空间变换、富有体验感的景观，打造了公园式景观体验区。中心水景广场拾阶而上，花团锦簇、流水潺潺，成为休憩、观光的好场所。庭院结合丰富的中国传统文化元素，如木屏风、"寿"字地雕、特色青砖、水缸等元素，形成步移景换、富有意境的自然生态庭院景观。经过尊贵的入口景观区，进入精雕细凿的样板庭院，便开始了一段赏心悦目的体验之旅。

The design of the park-like landscape experience area is presented with a Chinese landscape style. Through elements including screams, landscape wood platforms, rockeries, plantings, a landscape of spatial alternation, one step, one scene and rich experience feeling is formed. The stepped central waterscape plaza, surrounded by blooming flowers and babbling water, is a great place for relaxation and sight-seeing. In addition, the courtyard combines various traditional Chinese culture elements, including wood screen, sculpture in the shape of Chinese Character "Shou" (which means "longevity" in English), feature blue bricks and water vat, building a natural ecological courtyard landscape rich in artistic conception and one step, one scene. Once passing through the exalted entrance landscape area and entering into the delicately carved prototype courtyard, a delightful experience tour begins.

04 南昌万达城实景示范区庭院
05 南昌万达城实景示范区园路
06 南昌万达城实景示范区景观木栈道

DEMONSTRATION AREA OF WUXI WANDA CITY
无锡万达城实景示范区

开放时间 2014 / 08 / 23　　**OPENED ON** AUGUST 23 / 2014
项目位置 江苏 / 无锡　　　　**LOCATION** WUXI / JIANGSU PROVINCE
占地面积 2.2 公顷　　　　　　**LAND AREA** 2.2 HECTARES

PLANNING OF DEMONSTRATION AREA
示范区建筑规划

无锡万达城实景示范区占地2.2万平方米，实体样板间分别位于场地的东南角和西北角。示范区设计通过销售动线的组织，多维度展现万达住宅环境的卓越品质。

The demonstration area of Wuxi Wanda Cultural Tourism City occupies an area of 22,000 square meters. The prototype rooms are located at the southeast and northwest corners of the site. Through the organization of sales circulation, the design of the demonstration area reveals the excellent quality of the environment of Wanda residences in multiple dimensions.

03

01 无锡万达城实景示范区总平面图
02 无锡万达城实景示范区庭院
03 无锡万达城实景示范区水景
04 无锡万达城实景示范区入口景观

04

LANDSCAPE OF DEMONSTRATION AREA
示范区景观

中式的景观立意，取得建筑与景观藏露呼应、虚实相间的效果。掩映中的框景墙、漏窗、门洞，将若隐若现的山石飞瀑"借入"眼帘；进入庭院前的门楼，寄托了徜徉于城市山林之中的怡然心态；曲径通幽的后院、蜿蜒而过的溪涧，诠释了新古典园林的情怀。

The landscape design adopts Chinese style, achieving a coordination of hiding and showing and alternation of reality and virtuality. The enframed walls, ornamental perforated windows and door openings "borrow" the partly hidden and partly visible hill stones and waterfalls into view, the gate leading to the courtyard conveys the pleasant and contented attitude of wandering about urban hills and woods; the quiet backyard with winding paths and streams demonstrates the romanticism of new classical garden.

07

05 无锡万达城实景示范区水景
06 无锡万达城实景示范区景观庭院
07 无锡万达城实景示范区景观庭院
08 无锡万达城实景示范区庭院夜景

08

DEMONSTRATION AREA OF QINGDAO ORIENTAL CINEMA
青岛东方影都实景示范区

开放时间 2015 / 05 / 30　　OPENED ON MAY 30 / 2015
项目位置 山东 / 青岛　　　　LOCATION QINGDAO / SHANDONG PROVINCE
占地面积 2.7 公顷　　　　　 LAND AREA 2.7 HECTARES

PLANNING OF DEMONSTRATION AREA
示范区建筑规划

建筑采用富有流动感的空间结构形式，把服务区、展示区、体验区等多种功能区域与优美的室外风景统合起来。建筑强调社会参与性，通过流动的屋顶、纯净的幕墙和自由舒展的环境，让人们感受万达"东方影都"带来的轻松生活方式。

The building adopts a spatial structural style that has a strong sense of flow, combining the various function areas including the service area, the exhibition area and the experience area with the beautiful outdoor scenery. With an emphasis on social interaction, the building, through roofs with flowing lines, pure and clear curtain walls and a free and stretching environment, allows people to perceive the relaxed lifestyle brought by Wanda Oriental Cinema.

01 青岛东方影都实景示范区总平面图
02 青岛东方影都实景示范区建筑外立面
03 青岛东方影都实景示范区建筑设计手绘图

PART B CULTURAL TOURISM PROJECT
文旅项目

LANDSCAPE OF DEMONSTRATION AREA
示范区景观

景观设计的创意源自青岛的生态环境——沿岸景观、流动的海浪形态及闪烁的海边砾石。从这些自然意象中，寻找了"水、木、草、石"四种元素，呈现如海般自然生态的城市生活景观和如影般休闲的文化生活理念。绿化以当地的乡土树种为骨架，自然饱满的地形为依托，同时以景石进行有序的搭配，或开或合、疏密有致。徜徉其间，曲径通幽，偶然邂逅一片光影斑斓，绿意通幽，使得展示区若隐若现，宛若依傍在海边的"林间小屋"！

The landscape design idea originates from Qingdao's ecological environment-coastal landscape, flowing sea waves and sparkling seaside gravels. Four elements of "water, wood, grass and stone" are extracted from these natural images to demonstrate an ocean-like natural and ecological urban life landscape and a casual and movie-like cultural life philosophy. Based on natural and plump terrain, the greening adopts indigenous trees in orderly coordination with scenery stones, presenting alternated openness or enclosure with natural density. Walking inside, winding paths will lead you to secluded quiet place, where you shall unexpectedly meet multi-colored lights and shadows and green scenes, making the whole demonstration area look like a "wood cottage" by the seaside, partly hidden and partly visible.

CULTURAL TOURISM PROJECT
文旅项目

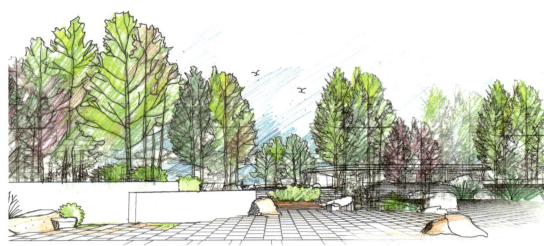

06

04 青岛东方影都实景示范区景观
05 青岛东方影都实景示范区水景
06 青岛东方影都实景示范区景观设计手绘效果图
07 青岛东方影都实景示范区景观木栈道

07

DEMONSTRATION AREA OF HEFEI WANDA CITY
合肥万达城实景示范区

开放时间 2013 / 11 / 02　　OPENED ON NOVEMBER 2 / 2013
项目位置 安徽 / 合肥　　　LOCATION HEFEI / ANHUI PROVINCE
占地面积 1.15 公顷　　　　LAND AREA 1.15 HECTARES

PLANNING OF DEMONSTRATION AREA
示范区建筑规划

示范区采用开放社区、三级物管，打造金街社区生活品质。人车分流、全景地库设计、社区景观大花园，营造出公园般居住环境。建筑采用欧式新古典风格，运用现代手法和工艺，将古典元素进行符号化抽象，在讲究风格、追求意境的同时，快速呈现完美效果。

The demonstration area is an open community with three-level property management, creating a quality life for golden street community. A park-like living environment is created through the design of separation of pedestrian and vehicles, the panoramic basement and the large community landscape garden. The building adopts European neoclassical style. Through the application of modern techniques and crafts, classical elements are symbolically abstracted to speedily present perfect effects while highlighting style and pursuing artistic conception.

02

CULTURAL TOURISM PROJECT
文旅项目

01 合肥万达城实景示范区建筑外立面
02 合肥万达城实景示范区总平面图
03 合肥万达城实景示范区景观
04 合肥万达城实景示范区景观
05 合肥万达城实景示范区水景

LANDSCAPE OF DEMONSTRATION AREA
示范区景观

以"水灵动,墨淡雅,茶闲韵"的徽派生活元素为设计理念,承容"巢湖文化",打造现代中式风格景观蓝本。坡地、树林、水系、草坪,自然灵动、曲径通幽,在呈现景观效果的同时,赋予不同空间相应的功能性,让业主参与其中,同时感受美与趣味!

Adopting Hui-style life elements of "flowing water, elegant Chinese ink and leisure tea-drinking" as the design concept and incorporating "Chaohu Culture", the demonstration area aims to create a modern Chinese-style landscape model. The sloping land, woods, waters and lawns are natural and dynamic, with winding paths leading to secluded quiet place, which not only demonstrate landscape effects but also endow corresponding functions to different spaces. Owners involved may experience its beauty and fun.

COMMERCIAL PROJECT
商业项目

WANDA
COMMERCIAL
PLANNING
2014

ABOUT PROPERTIES FOR SALE OF WANDA COMMERCIAL REAL ESTATE
说说万达商业地产的销售物业

万达商业地产设计中心南区设计部总经理　张东光

提到万达，大家往往想到的是"万达广场"、"商业综合体"或"万达影院"。对万达的销售物业，知道得不多、不全面。那么万达的销售物业是怎么回事呢？它在万达"商业帝国"中扮演的是什么角色呢？

销售物业是万达商业地产的"输血机"，有着与其他房地产公司销售物业不一样的特点。作为"输血机"，销售物业必须按时、保质地提供"血液"（资金），确保现金流。这个特点和承担的角色，决定了万达商业地产销售物业的特性。

要及时"输血"，就必须实现产品的快销。而要做到快销，就必须有能支持快销、能打动人的产品，这是重中之重。万达商业地产设计中心就是这类优质产品的创造者。

为配合万达销售类物业的战略定位，万达商业地产设计中心做了如下四个方面的工作。

一、优秀的卖场

销售道具是客户了解产品最直接的途径，也是客户对万达销售物业的第一印象。我们通过深入思考和认真探索，找到用优质卖场的效果和突出的品质吸引客户的办法——首先要体现万达产品的特点；其次要保证品质，又好又快；第三还要控制成本。为此。我们通过不断总结梳理，形成了优质的卖场模块及管理要点，坚持走模块化、标准化、产业化的道路。

二、产品的创新

产品的创新是保持产品生命力的法宝之一。我们一直不断地总结过往成熟的产品、研究市场的变化、了解人们生活习惯的变化，同时预判科技的发展对居住方式的影响，有的放矢地进行产品的优化和创新。这也是万达商业地产销售物业保持生命力，可以不断地打动客户的根本支撑。我们的"百变空间"、"安全屋"、"智能家居"等创新，无一例外成为客户津津乐道的话题。

When referring to Wanda, it is usually associated with "Wanda Plaza", "Commercial Complex" or "Wanda Cinema". Yet there is limited and incomplete perception of Wanda Properties for Sale. What is Wanda Properties for Sale on earth? And what is the role of it in Wanda "Commercial Empire"?

Served as the "blood transfusion machine", Properties for Sale has unique features compared with properties for sale in other real estate companies, and must provide blood (capital) on time with high quality to ensure the cash flow. The feature and the role as such endow Properties for Sale of Wanda Commercial Real Estate with distinctive quality.

To realize timely blood transfusion, fast selling of product is a must. The top priority is that fast selling must be supported by impressive products, which are exactly created by Wanda Commercial Real Estate Design Center.

To cooperate with the strategic positioning of Wanda properties for sale, Wanda Commercial Real Estate Design Center has done the four aspects of work as below:

I. QUALITY SALES SPACE

Sales props are the most direct way for clients to understand the product and the first impression of clients on Wanda properties for sale as well. Through in-depth thinking and careful exploration, we turn to the approach of attracting clients via quality effect and outstanding quality. Firstly, the features of Wanda products shall be presented; secondly, quality shall be guaranteed, being better while faster; thirdly, the cost shall be under control. To achieve this, we have formed high-quality sales space module and management key points through unceasingly summarization and arrangement, and walk on the path of standardization, modularization and industrialization.

II. PRODUCT INNOVATION

Product innovation is one of the magic weapons to keep the vitality of product. Through constantly summarizing the previous mature products, studying the market changes, perceiving the variation of people's living habits, and predicting the influence of development of science and technology on the way of living, we apply targeted product optimization and innovation. This is the base for Properties for Sale of Wanda Commercial Real Estate to keep vitality and continually impress clients. It is no surprise that clients take delight in talking about our innovations such as "Flexible Space", "Safe House" and "Smart Home".

三、研发及专利

产品的创新离不开研发的支持,万达商业地产销售物业必须不断有新的卖点来打动客户,同时要更有效地节省成本。开源节流是保证利润的根本。

为了更好地从产品本身开源节流,万达设计人一直在不断地努力。安全屋的落地为营销创造了新的卖点,极大地促进销售;"景观部品库"的落地打响了万达产业化的第一枪;"地库的研发创新"增加了车位,减少了地库面积,优化了土方,改善了地下大堂的品质;"百变空间"为小户型的利用提供了无限的可能,同时取得了5项专利。

四、收获及奖项

通过万达设计人的努力,万达销售物业的产品得到了客户的认可,开盘既售罄的现象在2014年不乐观的地产形势下频频出现。

客户的认可是一方面,说明万达的产品接地气;另一方面,万达的产品也是"高大上"的。在由中国建筑学会主办的"2014全国人居经典建筑规划设计方案竞赛"中,万达成为最大的赢家,5个项目获得8项大奖。既高大上又接地气的万达产品,自然成为市场的宠儿。

新的一年即将来临,房地产市场仍不明朗,但我们认为设计的价值因逆势而存在。万达商业地产设计中心将进一步整合资源,全面提升综合能力,为销售助力,谱写辉煌的新篇章。

III. TECHNOLOGY R&D AND PATENT

Innovation on products can't be divorced from the support of technology R&D. Properties for Sale of Wanda Commercial Real Estate keep creating new selling points to impress customers and meanwhile control the cost in a more efficient manner. To put it simply, increasing income and reducing expenditure is the root of profit.

Then, how to achieve this from the product itself. Wanda designers always keep trying. The establishment of "Safe House", for example, creates new selling point for marketing and greatly promotes the sales; the implementation of "Landscape Ornaments Library" fires the first shot of industrialization of Wanda; "R&D innovation in basement" increases the parking space, reduces the basement area, optimizes the earthwork, and improves the quality of underground lobby; "Flexible Space" provides more use possibilities for small-sized house type and wins 5 patents.

IV. PARTICIPATING IN COMPETITIONS FOR HONOR

Through the efforts of Wanda Designers, products of Wanda Properties for Sale have won the acceptance of the clients. The spectacle of "Sold out on Opening" is frequently witnessed in 2014 against the depressing industry trend.

Approval of the clients is one reason, showing that Wanda products are widely accepted by people, and high-end Wanda products are the other reason. In 2014 National Human Settlement Classical Architecture Planning Competition sponsored by the Architectural Society of China, for example, Wanda, with 5 projects winning 8 awards in total, becomes the biggest winner. Benefiting from the "down-to-earth" and "high-end" Wanda products, Wanda is sure to be favored by the market.

In the forthcoming year, real estate market still faces a not so bright condition, but we believe that the value of design can be brightened by adverse situations. Wanda Commercial Real Estate Design Center will further integrate resources and improve comprehensive ability to promote sales and continue its gloriously new chapter.

2014 WANDA COMMERCIAL PLANNING
万达商业规划——销售类物业

优秀项目 01

NANJING JIANGNING WANDA MANSION
南京江宁万达公馆

入伙时间 2014 / 09 / 27	OPENED ON SEPTEMBER 27 / 2014
项目位置 江苏 / 南京	LOCATION NANJING / JIANGSU PROVINCE
占地面积 10.11 公顷	LAND AREA 10.11 HECTARES
建筑面积 53.34 万平方米	FLOOR AREA 533,400m²

PLANNING OF MANSION
公馆规划

项目位于南京市江宁区竹山路两侧，用地以竹山路为界分为东西两部分，西侧地块为购物中心、商业街及写字楼，东侧地块为住宅、底商及五星级酒店。

Located on both sides of Zhushan Road in Jiangning District of Nanjing, the site of the Mansion is divided into two parts by the Zhushan Road, wherein the west part holds a shopping center, a commercial street and an office building, while the east part accommodates residences, stores at the bottom and a five-star hotel.

01 南京江宁万达公馆总平面图
02 南京江宁万达公馆建筑夜景
03 南京江宁万达公馆入口
04 南京江宁万达公馆户型图

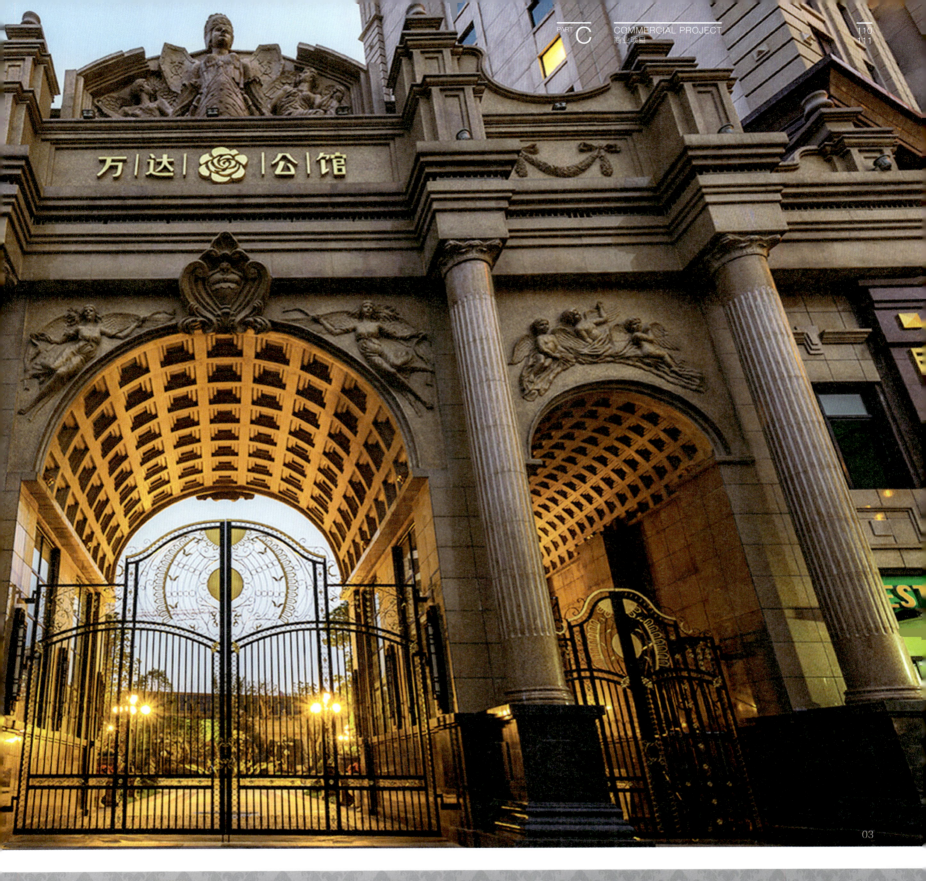

BUILDING OF MANSION
公馆建筑

采用对称的法式古典风格，庄重大方，富有气势。户型设计强调规整方正，分区明确。采用生活流线、家政流线"双流线"设计，各个功能既独立又融合，互不打扰也使用便捷。主要卧房均采用套间设计，超大尺寸的主卧室和客厅，突显高级住宅的卓越品质。

The grand and dignified building adopts the symmetrical French classical style. The design of the house layout attaches importance to order, and clear function division. Adopting a "dual-streamline" design covering living and housekeeping, function areas are separate yet integrated, each area being user-friendly without interference to each other. With the main bedrooms all designed as suite, the super large main bedroom and living room highlight the outstanding quality of elite residence.

LANDSCAPE OF MANSION
公馆景观

法式风格，宁静温馨。通过时间轴线和空间轴线两条线索，选取了最具代表性的德洛卡德罗花园、诗歌音乐广场、普罗旺斯、第戎小城、枫丹白露城堡五个名园和地域元素，体现法式的经典与永恒，洋溢着浪漫情调。

The mansion, which adopts French style, is quiet and sweet. Following the time axis and space axis, five representative celebrated gardens (the Trocadero Garden, the Poetry Music Square, Provence, the Town of Dijon and Fontainebleau Castle) and regional characteristics are selected to reflect the classic and timeless French style and to show a romantic sentiment.

05 南京江宁万达公馆景观
06 南京江宁万达公馆景观花坛
07 南京江宁万达公馆景观雕塑
08 南京江宁万达公馆景观喷泉

PART C | COMMERCIAL PROJECT
商业项目

07

08

09

09 南京江宁万达公馆客厅内装
10 南京江宁万达公馆户型图
11 南京江宁万达公馆餐厅内装
12 南京江宁万达公馆卧室内装
13 南京江宁万达公馆客厅内装

INTERIOR OF MANSION
公馆内装

内装延续室外法式古典风格。罗马柱、雕花顶、大理石地面，结合室内空间，体现欧式古典风格的精致奢华。细腻的柱头雕花、精美的床头木雕配以金银设色，把精致推向了新的高度。

The interior design inherits the French classical style of the exterior design. Combing the interior space, the Roman columns, the ceiling with carvings and the marble floor reflect the delicacy and luxury of European classical style. The gold and silver colored exquisite carvings on the column head and the fine wood carvings in the bedside together push the delicacy of the interior design to a new level.

2014 WANDA COMMERCIAL PLANNING
2014 万达商业规划——销售类物业

XI'AN DAMING PALACE WANDA MANSION
西安大明宫万达公馆

入伙时间 2014 / 09 / 30	OPENED ON SEPTEMBER 30 / 2014
项目位置 陕西 / 西安	LOCATION XI'AN / SHAANXI PROVINCE
占地面积 4.63 公顷	LAND AREA 4.63 HECTARES
建筑面积 25.41 万平方米	FLOOR AREA 254,100m²

PLANNING OF MANSION
公馆规划

西安大明宫万达公馆位于西安市未央区大明宫国家遗址公园北侧，与商业中心、室外商业街、精装SOHO、5A写字楼、圣马丁风情商业街等业态融于一体，实现了商业娱乐、商务办公、高端居住、餐饮休闲的多功能复合。

Located on the north side of the Daming Palace National Heritage Park in the Weiyang District of Xi'an, the Xi'an Daming Palace Wanda Mansion is blended with various business types including commercial center, outdoor commercial street, well-decorated SOHO, Grade 5A office building and Saint Martin style commercial street, creating a multifunctional complex for commercial entertainment, business office, high-end inhabitancy, catering and leisure.

01 西安大明宫万达公馆入口
02 西安大明宫万达公馆景观

BUILDING OF MANSION
公馆建筑

外立面采用法式古典风格，立面造型对称、强调理性，立面线条鲜明、凹凸有致，色彩稳重大气。孟莎式屋顶，造型各异的老虎窗、廊柱、雕花、线条，处处体现富贵典雅的风格。

The exterior facade adopts French-style classic style, boasting of symmetric and rational modeling, clear-cut and irregular lines and dignified and grand color. A luxurious and elegant style is reflected all over from the mansard roof, dormers of different shapes, aisle columns, carvings and lines.

03 西安大明宫万达公馆住宅入口
04 西安大明宫万达公馆室外步行街
05 西安大明宫万达公馆住宅入口

06

06 西安大明宫万达公馆景观雕塑
07 西安大明宫万达公馆景观喷泉
08 西安大明宫万达公馆景观小品
09 西安大明宫万达公馆景观水池
10 西安大明宫万达公馆景观设计手绘剖面稿

07

08

LANDSCAPE OF MANSION
公馆景观

通过对场地性质、空间及居住区景观趋势的分析与总结,引入法式古典园林序列感强、层层渐进的轴线景观,塑造了新古典美学原则下的纵轴、端点、界面、横轴、节点的景观结构,打造出纯粹高贵的法式园林。

Based on an analysis and summarization of the nature of the site, the space and the residential landscape tendency, a axis landscape with a strong sense of order and layered progression that are commonly seen in French-style classical gardens is brought in to build a pure and noble French-style garden, creating a landscape structure with vertical axes, end points, interfaces, horizontal axes and nodes under the new-classical aesthetic principle.

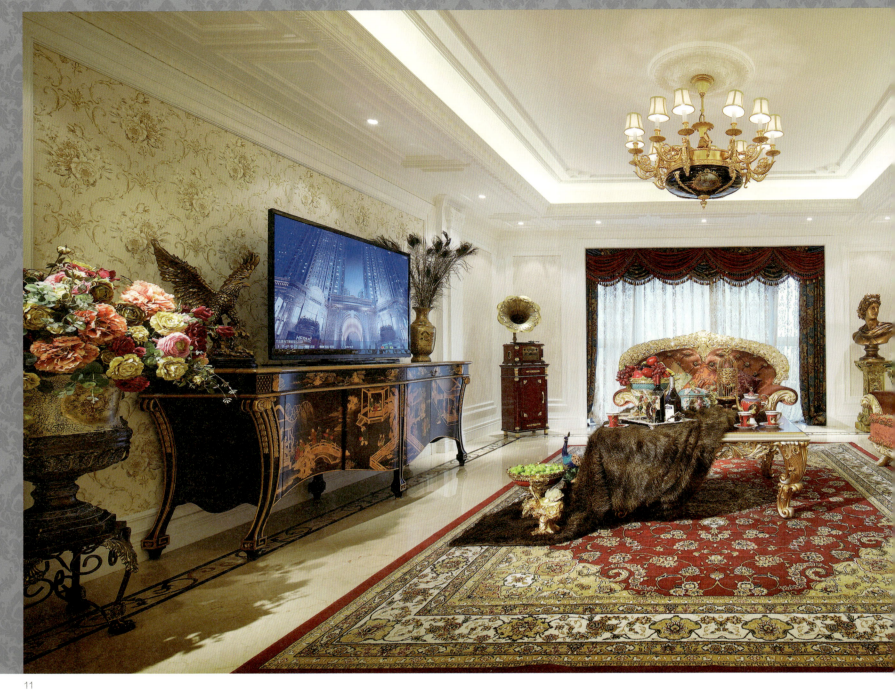

INTERIOR OF MANSION
公馆内装

室内以米色为主基调，辅以香槟金色作为细节装饰，重点部位以活跃的对比色加以突出，其余部分以浅色调协调统一。顶棚造型精致考究，灯具、雕花、金箔，体现了细节之美。

The interior design adopts light beige as the main color, accompanied by champagne gold for detail decoration. The key parts are highlighted with active contrasting colors, and the remaining parts are coordinated and unified with light colors. The modeling of the ceiling is delicate and exquisite, and the lamps, carvings and gold foil all reflect beauty in details.

11 西安大明宫万达公馆客厅内装
12 西安大明宫万达公馆户型图
13 西安大明宫万达公馆内装
14 西安大明宫万达公馆卧室内装

PART **C** COMMERCIAL PROJECT
商业项目

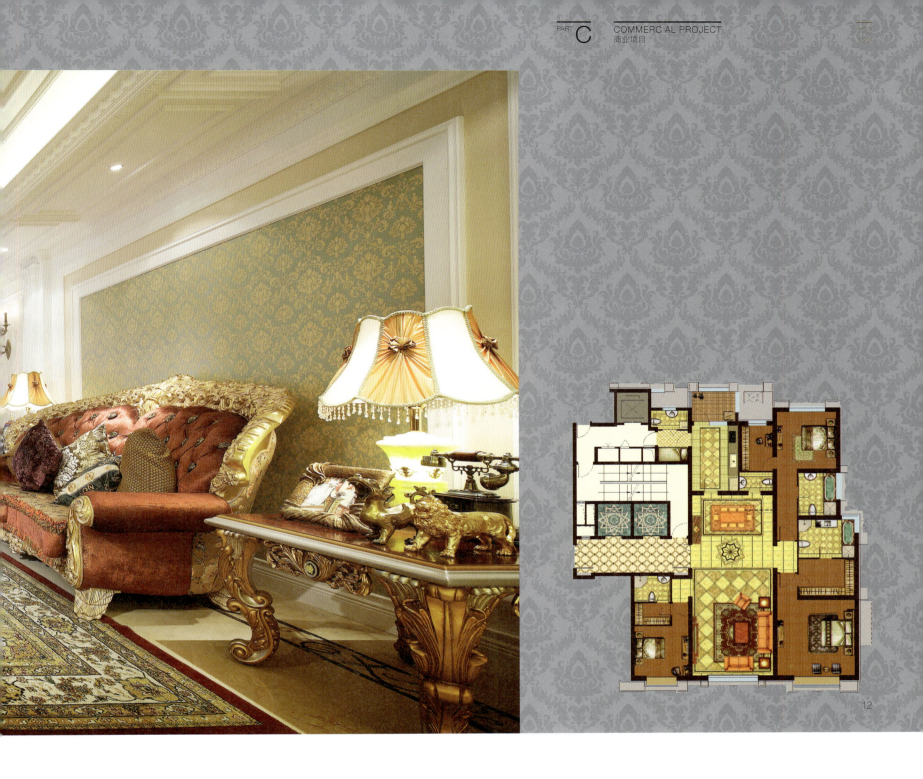

2014 WANDA COMMERCIAL PLANNING
2014 万达商业规划——销售类物业

优秀项目 03

FUQING WANDA PALACE
福清万达华府

入伙时间 2014 / 12 / 30	**OPENED ON** DECEMBER 30 / 2014
项目位置 福建 / 福州	**LOCATION** FUZHOU / FUJIAN PROVINCE
占地面积 8.01 公顷	**LAND AREA** 8.01 HECTARES
建筑面积 41.16 万平方米	**FLOOR AREA** 411,600m²

B区总平面图

PLANNING OF PALACE
华府规划

福清万达广场位于福清市清昌大道北侧，西环路东侧，清宏路南侧，涵盖购物中心、室外步行街、5A甲级写字楼、高档住宅四大业态，集购物、休闲、娱乐、商务、居住等多种城市功能，建成后成为福清体量最大、商业档次最高的商业中心和城市地标建筑。

Located on the north side of the Qingchang Road, the east side of the Xihuan Road and the south side of the Qinghong Road in Fuqing, the Fuqing Wanda Plaza contains four business types of shopping center, exterior pedestrian street, Grade 5A office building and high-end residence, integrating multiple city functions including shopping, leisure, entertainment, business and inhabitancy. Upon completion, it becomes the largest commercial center and city landmark building of the highest commercial level.

01 福清万达华府总平面图
02 福清万达华府建筑设计手绘效果图
03 福清万达华府建筑外立面

PART **C** | COMMERCIAL PROJECT
商业项目

BUILDING OF PALACE
华府建筑

建筑立面用竖向窗带组织，纵向颜色深浅分明，整体轮廓高耸挺拔，形成高贵内敛的艺术气质。构图强调对称，营造古典秩序感的同时，以简约、洗练、纯粹的风格，使居住者情感回归宁静与自然，彰显了精品住宅的尊贵品位。

The facade of the building adopts vertical window straps, with vertically clear-cut dark and light color distribution. Being tall and upright, the whole outline forms an elegant and restrained artistic temperament. The composition of the building stresses symmetry to create a classical order impression, and emotionally, it brings the residents back to peace and nature by following a contracted, concise and pure style, manifesting the exalted status of the quality residence.

LANDSCAPE OF PALACE
华府景观

景观设计采用自然风景园的设计手法，将微地形、水、植物与园林建筑、小品等结合，创造自然与艺术相得益彰的小区环境。结合地形变化营造聚会、休息、健身、游赏等多种形态的功能场所，满足小区不同人群的多种需求，创造丰富的生活体验。

Adopting the design method of natural scenery garden, the landscape design combines micro-topography, water, plant, garden architecture and architectural pieces, creating a community environment where nature and art complement each other. Combined with topographical changes, function occasions for various purposes including holding parties, having a rest, keeping fit and visiting and touring have been built, thus satisfying multiple demands of different groups of people in the community and creating a rich life experience.

04 福清万达华府入口夜景
05 福清万达华府住宅入口
06 福清万达华府景观设计手绘效果图
07 福清万达华府景观亭
08 福清万达华府景观

INTERIOR OF PALACE
华府内装

内装设计追求奢华、稳重的风格,于古典柱式、顶棚造型、描金线条、大理石拼花地面等细节之处,营造欧式古典氛围。细致的雕花、精雕细选的配色,彰显主人对生活品位的追求。

The interior design pursues a luxurious yet steady style, creating a European-style classical atmosphere through details including classical column types, ceiling modeling, outline in gold and marble mosaic floors. Exquisite carvings and well-selected color combination manifest the owner's pursuit for quality lifestyle.

PART C　COMMERCIAL PROJECT
商业项目

11

12

09 福清万达华府户型图
10 福清万达华府客厅内装
11 福清万达华府门厅内装
12 福清万达华府卧室内装

2014 WANDA COMMERCIAL PLANNING
万达商业规划——销售类物业

售楼处 01

SALES OFFICE OF DALIAN JINGKAI WANDA PLAZA
大连经开万达广场售楼处

开放时间 2014 / 04 / 26　**OPENED ON** APRIL 26 / 2014
项目位置 辽宁 / 大连　**LOCATION** DALIAN / LIAONING PROVINCE
占地面积 0.14 公顷　**LAND AREA** 0.14 HECTARE
建筑面积 2300 平方米　**FLOOR AREA** 2,300m²

01 大连经开万达广场售楼处建筑外立面
02 大连经开万达广场售楼处总平面图

BUILDING OF SALES OFFICE
售楼处建筑

运用金属、高透玻璃、彩釉玻璃等材料，通过体量穿插、泛光照明、色彩对比，创作出现代风格的作品。与项目所处的大连开发区的规划理念吻合，定位准确。

With the application of metal, transparent glass and enameled glass, a piece of modern style work is created by means of alternative faceted volume, floodlighting and color contrast. The accurately-positioned building is in consistent with the planning concept of Dalian Development Zone where the project locates.

02

04

03 大连经开万达广场售楼处大厅内装
04 大连经开万达广场售楼处沙盘

INTERIOR OF SALES OFFICE
售楼处内装

内装设计着重体现"文明、社交、魅力"的理念。总服务台、接待区、沙盘区、区位图等功能分区明确，动线合理。菱形元素贯穿顶棚、墙面、地面，整个空间浑然一体、气势恢宏。海洋元素的喷绘玻璃、鱼群吊灯，明确了地域特点，起到"画龙点睛"的作用。

The interior design stresses the idea of "civilization, social contact and charm". The function division of reception desk, reception area, building model and location map is clear and definite with well-organized circulation. Rhombus elements run all over the ceiling, the walls and the floor, making the whole space look unified and imposing. As a finishing touch, the ocean-themed painted glass and the chandelier decorated with fish elements highlight the regional characteristics of Dalian.

05

06

07

2014 WANDA COMMERCIAL PLANNING
万达商业规划——销售类物业

售楼处 02

SALES OFFICE OF PANJIN WANDA PLAZA
盘锦万达广场售楼处

开放时间 2014 / 06 / 28	OPENED ON JUNE 28 / 2014
项目位置 辽宁 / 盘锦	LOCATION PANJIN / LIAONING PROVINCE
占地面积 3200 平方米	LAND AREA 3,200m²
建筑面积 1500 平方米	FLOOR AREA 1,500m²

BUILDING OF SALES OFFICE
售楼处建筑

建筑外立面采用Art Deco风格，强调古典秩序感，运用虚实对比的手法，营造出优雅高贵的气质。错落有致的屋顶、疏密得当的竖向线条，使得售楼处建筑富有韵律感。

The facade of the building adopts Art Deco style, which stresses classical order impression and creates an elegant and graceful temperament by means of virtual-real comparison. The well-arranged roof and the well-proportioned vertical lines render the sales office building a strong sense of rhythm.

01 盘锦万达广场售楼处平面图
02 盘锦万达广场售楼处建筑外景
03 盘锦万达广场售楼处建筑外景

INTERIOR OF SALES OFFICE
售楼处内装

内装设计延续外立面风格，将盘锦当地特有的红海滩、芦苇荡、麦穗等自然元素，抽象为装饰元素，运用在空间造型及色彩设计中。整体空间开阔、功能空间排列有序、色调柔和，气氛热烈，为项目的营销提供了有力的支撑。

The interior design follows the style of the facade. Natural elements with unique and regional characteristics including red beach, reed marshes and ears of wheat are abstracted as decorative elements and applied to space modeling and color design. The sale of the project is strongly promoted by the wholly open space and the well-organized functional space and gentle yet impassioned colors.

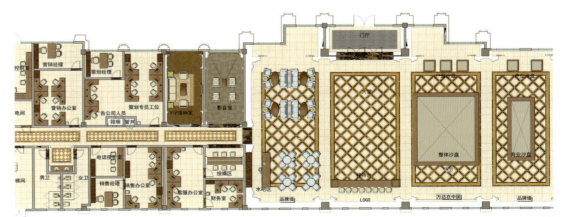

04 盘锦万达广场售楼处大厅内装
05 盘锦万达广场售楼处洽谈区内装
06 盘锦万达广场售楼处平面图（局部）

2014 WANDA COMMERCIAL PLANNING
万达商业规划——销售类物业

SALES OFFICE OF DONGGUAN HUMEN WANDA PLAZA
东莞虎门万达广场售楼处

开放时间 2014 / 08 / 24　　**OPENED ON** AUGUST 24/ 2014
项目位置 广东 / 东莞　　　　**LOCATION** DONGGUAN / GUANGDONG PROVINCE
占地面积 886 平方米　　　　**LAND AREA** 886m²
建筑面积 1164 平方米　　　　**FLOOR AREA** 1,164m²

01 东莞虎门万达广场售楼处平面图
02 东莞虎门万达广场售楼处外立面
03 东莞虎门万达广场售楼处入口

BUILDING OF SALES OFFICE
售楼处建筑

外立面高耸、挺拔，巨大的穹顶、庄严的柱式，完美诠释古典建筑之美。主厅平面通过一系列既分隔又连通的空间互相衔接、穿插，完成功能分区，使人在行进中感受到丰富的空间变化及良好的卖场氛围。

The beauty of classical architecture is perfectly demonstrated by the towering and straight facade, the huge dome and the grand columns. The well functionally divided main hall is linked and interspersed by a series of both separate and connected spaces, enabling people to witness a rich space variation and a desirable sales atmosphere while walking around.

01

INTERIOR OF SALES OFFICE
售楼处内装

布局突出轴线对称，细节突出考究的工艺线条，色彩柔和优雅，动线通畅合理。大面积墙面以高档石材为主，点缀深色皮革及玫瑰金包边，地面拼花铺、顶棚灯槽配合巨大的水晶吊灯，使整体空间氛围主重、热烈。

The layout highlights axis symmetry; the details reveal exquisite lines; the color is gentle and graceful; the circulation is smooth and rational. Adopting quality stone as the major material for most of the walls, embellished with dark leather and rose-gold margins, and combined with the mosaic floor and the huge crystal chandelier installed in the light trough on the ceiling, the whole space presents a solemn and enthusiastic atmosphere.

04 东莞虎门万达广场售楼处前台
05 东莞虎门万达广场售楼处洽谈区内装
06 东莞虎门万达广场售楼处平面图
07 东莞虎门万达广场售楼处大厅内装

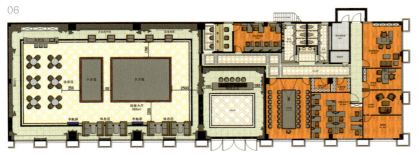

LANDSCAPE OF SALES OFFICE
售楼处景观

景观设计突出轴线的对称性，对售楼处前广场进行了合理的功能分区。阵列的景观灯柱及道旗，增强了仪式感、尊贵感和商业氛围。休闲活动区营造轻松氛围吸引人流驻足，最大限度地利用广场空间进行宣传展示活动。

The landscape design highlights axis symmetry, based on which the square in front of the sales office enjoys a reasonable function division. The arrays of landscape lamp posts and banners enhance the ritual sense, prestige perception and commercial atmosphere. The relaxed atmosphere of the leisure activity area helps to attract people flow, and the square space can be used to the greatest advantage for promotion and exhibition activities.

08 东莞虎门万达广场售楼处景观广场
09 东莞虎门万达广场售楼处景观伞座
10 东莞虎门万达广场售楼处景观设计效果图
11 东莞虎门万达广场售楼处景观绿化

2014 WANDA COMMERCIAL PLANNING
万达商业规划——销售类物业

SALES OFFICE OF WANDA REIGN CHENGDU
成都瑞华酒店售楼处

开放时间 2014 / 09 / 01	OPENED ON SEPTEMBER 1 / 2014
项目位置 四川 / 成都	LOCATION CHENGDU / SICHUAN PROVINCE
占地面积 550 平方米	LAND AREA 550m²
建筑面积 440 平方米	FLOOR AREA 440m²

01 成都瑞华酒店售楼处外立面
02 成都瑞华酒店售楼处建筑设计手绘图
03 成都瑞华酒店售楼处外立面
04 成都瑞华酒店售楼处总平面图

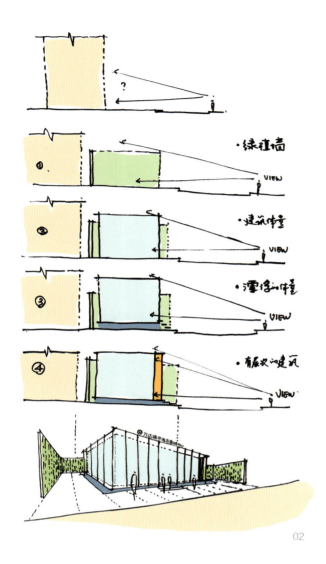

BUILDING OF SALES OFFICE
售楼处建筑

采用与酒店建筑协调的水平延展体量 以形态纤细的柱式与体量壮观的酒店立面呼应，体现了古典建筑比例、尺度的和谐统一之美。

The sales office building adopts a horizontally extended volume that is in coordination with the hotel building and echoes the magnificent hotel facade with slender and fine columns, reflecting the harmonious and unified beauty in the proportion and scale of classical architecture.

05 成都瑞华酒店售楼处夜景

PART C COMMERCIAL PROJECT
商业项目

07

08

06 成都瑞华酒店售楼处大厅内装
07 成都瑞华酒店售楼处沙盘
08 成都瑞华酒店售楼处洽谈区内装
09 成都瑞华酒店售楼处平面图

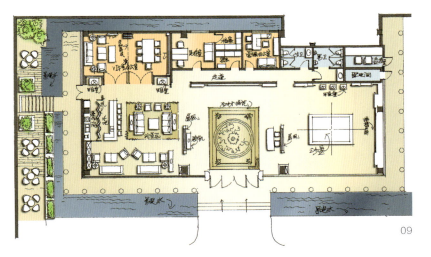

09

INTERIOR OF SALES OFFICE
售楼处内装

入口对景选用淡彩山水画石材拼图，突出背景墙的高阔气势。企业与项目展示区以高清LED巨屏动态地展现万达集团在行业的领先地位及国际化发展趋势。洽谈区重点营造奢华酒店环境，呼应顶级酒店的定位。

The opposite scenes at the entrance are pieced together with stones painted with light-colored mountains-and-waters, highlighting the tall and broad momentum of the background wall. At the enterprise and project exhibition area, a big high-definition LED screen dynamically exhibits Wanda Group's leading position n the industry and its international development trend. At the negotiation area, a luxurious hotel environment is created to echo with the positioning of Wanda Reign as a top hotel.

10 成都瑞华酒店售楼处景观花园
11 成都瑞华酒店售楼处景观设计手绘图
12 成都瑞华酒店售楼处水景

LANDSCAPE OF SALES OFFICE
售楼处景观

周边设计矩形镜面水池,用水景"托起"主体建筑,建筑在水中形成倒影,使外立面肌理在水中得以延续。后庭院闹中取静,营造适宜的商务洽谈环境。

Rectangle mirror pools are designed around to make the waterscape a foil of the main building. With inverted reflection in water, the facade texture of the building is extended in water. The backyard, being quiet in a noisy neighborhood, is a suitable place for business negotiation.

2014 WANDA COMMERCIAL PLANNING
万达商业规划——销售类物业

售楼处
05

SALES OFFICE OF MEIZHOU WANDA PLAZA
梅州万达广场售楼处

开放时间 2014 / 08 / 30　**OPENED ON** AUGUST 30 / 2014
项目位置 广东 / 梅州　**LOCATION** MEIZHOU / GUANGDONG PROVINCE
占地面积 0.61 公顷　**LAND AREA** 0.61 HECTARE
建筑面积 1900 平方米　**FLOOR AREA** 1,900m²

BUILDING OF SALES OFFICE
售楼处建筑

着重体现传统客家建筑的形态——长方形平面布局，功能流线竖向串联，达到"堂字屋"的"九井十八厅"的效果。建筑立面主要由铝板和玻璃组成，顶部以深灰色檐口再现客家建筑的灰瓦屋顶，体现了对传统建筑的传承。

The building focuses on the representation of the form of traditional Hakka architecture, which has a rectangle plane layout and vertically connected function flow, achieving the "Nine Wells and Eighteen Halls" effect of "Hakka House". The facade of the building mainly consists of aluminum plate and glass. The dark-gray cornices on the roof aim to represent the gray tiled roof of Hakka buildings, which reflect an inheritance to traditional architecture.

01

02

01 梅州万达广场售楼处外立面
02 梅州万达广场售楼处总平面图
03 梅州万达广场售楼处外立面
04 梅州万达广场售楼处建筑设计手绘图

INTERIOR OF SALES OFFICE
售楼处内装

采用现代新古典风格，以传统文化意象营造亲切的感观享受。顶棚覆盖充满活力的艺术空间——大型吊灯搭配香槟金线条，使光影、线条、空间比例和谐又有变化。

The interior design adopts new-classical style, creating an amiable perceptional feast with traditional cultural images. The vital art space under the ceiling – a huge chandelier matched with champagne gold lines, makes the light and shadows, lines and space scale all harmonious and varied.

07

05 梅州万达广场售楼处大厅内装
06 梅州万达广场售楼处内装设计手绘效果图
07 梅州万达广场售楼处平面图
08 梅州万达广场售楼处洽谈区内装

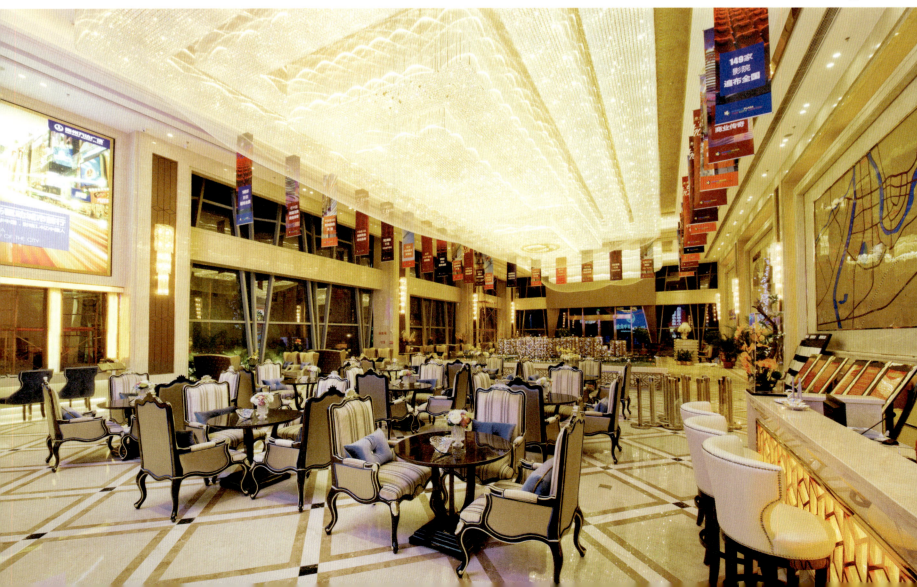

08

LANDSCAPE OF SALES OFFICE
售楼处景观

景观设计以"出水远航,客满天下"为主题。在景观周边布置镜面水池,池中设有层级跌水。主体建筑在怒放的花海、热烈的道旗映衬之下,犹如航行的巨轮,寓意"梅州万达"落户梅州,在商海中乘风破浪驶向辉煌。

The landscape design takes "Oceangoing Voyage in a Hakka World" as the theme. Mirror surface pools are laid out around the landscape with tiered waterfalls set inside the pool. The main building looks like a large sailing ship against the background of blooming flowers and lively banners, implying that "Meizhou Wanda" shall sail through winds and waves on its way towards glory.

09 梅州万达广场售楼处景观
10 梅州万达广场售楼处景观
11 梅州万达广场售楼处景观
12 梅州万达广场售楼处景观

09

10

11

2014 万达商业规划——销售类物业

售楼处
06

SALES OFFICE OF SHIYAN WANDA PLAZA
十堰万达广场售楼处

开放时间 2014 / 08 / 31	**OPENED ON** AUGUST 31 / 2014
项目位置 湖北 / 十堰	**LOCATION** SHIYAN / HUBEI PROVINCE
占地面积 1482 平方米	**LAND AREA** 1,482m²
建筑面积 2480 平方米	**FLOOR AREA** 2,480m²

BUILDING OF SALES OFFICE
售楼处建筑

整体布局紧密结合地形，围绕"仙山、秀水、汽车城"的元素整合设计。不规则线条自然地将多边形加以分隔；而这种分隔契合了内部的使用功能，使之虚实结合得恰到好处。建筑以不规则的折线、高起的门头进行诠释，呼应了汽车城的工业之美。

With the overall layout closely adjusting to terrain, the integrated design centers on the elements of "Celestial Mountain, Elegant Water and Motor City". The irregular lines divide the polygonal space naturally to accord with the internal use function of the space, making a perfect combination of vitality and reality. The industrial beauty of the motor city is echoed by the irregular broken lines and the raised gate of the building.

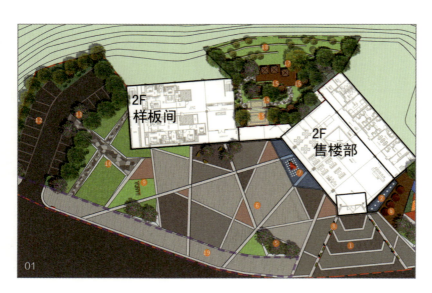

01 十堰万达广场售楼处总平面图
02 十堰万达广场售楼处鸟瞰图

INTERIOR OF SALES OFFICE
售楼处内装

内装设计结合建筑布局,将场地的"外"与"内"、"动"与"静"进行了充分"阐述",吊顶和地面的拼花图案延续外幕墙不规则折线,使室内外的设计风格融为一体。

The interior design is combined with the layout of the building, fully "interpreting" the "inside" and "outside" as well as the "dynamic" and "stillness" of the compound. The styles of interior and exterior designs have been integrated with the mosaic patterns on the ceiling and floor being consistent with the irregular broken lines on the external curtain wall.

03 十堰万达广场售楼处大厅内装
04 十堰万达广场售楼处景观小径
05 十堰万达广场售楼处景观小品

LANDSCAPE OF SALES OFFICE
售楼处景观

景观延续当地自然环境元素，与山体融为一体。提取建筑立面几何元素，用于地面肌理、绿化、水景；选用当地特色毛竹，结合场地特有的山体形态，营造小空间、大景观的效果。

The landscape adopts local natural environment elements as well and is integrated with the mountains. Geometrical elements extracted from building facade are applied to floor texture, greening and waterscape. Meanwhile, with the utilization of local moso bamboo and in combination with the unique mountainous terrain of the compound, a macro-landscape effect is achieved in a micro-space.

2014 WANDA COMMERCIAL PLANNING
万达商业规划——销售类物业

售楼处 07

SALES OFFICE OF NANNING WANDA MALL
南宁万达茂售楼处

开放时间 2014 / 09 / 06　**OPENED ON** SEPTEMBER 6 / 2014
项目位置 广西 / 南宁　**LOCATION** NANNING / GUANGXI ZHUANG AUTONOMOUS REGION
占地面积 1.6 公顷　**LAND AREA** 1.6 HECTARES
建筑面积 3300 平方米　**FLOOR AREA** 3,300m²

BUILDING OF SALES OFFICE
售楼处建筑

建筑设计灵感来源于广西少数民族的干阑式建筑——风雨桥、鼓楼。建筑整体如风雨桥一样舒展、主次分明、层次丰富。立面上的窗花及墙面纹饰，采用了当地最能体现广西少数民族历史文化风貌的铜鼓、壮锦。建筑细部工艺精美、结构严谨。

The design inspiration of the building derives from the stilt style architecture of minority nationalities in Guangxi, i.e. the Shelter Bridge and Drum-Tower. The whole building, like the Shelter Bridge, is stretching, clear in priorities and rich in layers. The window grille and wall ornamentation on the facade adopt local bronze drum and Zhuang brocade which can best reflect the historical and cultural characteristics of minority nationalities in Guangxi. The architectural details are exquisitely made and well structured.

01 南宁万达茂售楼处总平面图
02 南宁万达茂售楼处外立面
03 南宁万达茂售楼处外立面

02

03

入口广场上的LOGO柱、铜鼓雕塑、音乐喷泉、阵列花钵，一一对应建筑轴线。"壮锦"纹样的铺装和孔雀叠泉、仿真大铜鼓，结合色彩变幻的灯光突出地域文化特色，营造欢快愉悦的销售氛围。

The Logo column, bronze drum sculpture, music fountain and arrayed flowerpots in the entrance square are all in correspondence with the building axis. The pavement with decorative patterns of "Zhuang Brocade", the peacock cascading fountains and the huge simulation bronze drum, combined with changeable and colorful lights, highlight the characteristics of local culture and create a joyful and lively sales atmosphere.

04 南宁万达茂售楼处入口
05 南宁万达茂售楼处夜景
06 南宁万达茂售楼处入口

07 南宁万达茂售楼处洽谈区内装
08 南宁万达茂售楼处接待台
09 南宁万达茂售楼处会客区内装
10 南宁万达茂售楼处大厅内装

INTERIOR OF SALES OFFICE
售楼处内装

内装设计从南宁传统文化中提炼元素与灵感，加以重新解读——门厅两侧高达6米、运用多种石材拼织而成的"壮锦"图案，让人赞叹之余平添亲切之感；前台的铜鼓，沙盘正上方"朱槿"形态的水晶大吊灯，都堪称传神之作。

The interior design extracts elements and inspirations from traditional Nanning culture for re-interpretation. The 6 meters high pattern of "Zhuang Brocade" on both sides of the hallway pieced together with various types of stones not only amazes people but also makes people feel cordial; the bronze drum at the reception and the "China Rose" shaped chandelier above the building model are all vivid and impressive pieces of work.

LANDSCAPE OF SALES OFFICE
售楼处景观

"院式"空间，选择东南亚文化符号，色彩搭配温暖自然。休闲景观设施及情景式度假的景观元素——粗犷石材、佛教色彩、雕塑、小品、图腾柱、风情泰式凉亭、蓝色水系、亚热带植栽等——充分展现了亚热带风情景观。

The landscape adopts "court-style" space with Southeast Asia culture symbols and warm and natural colors. The recreation landscape facilities and situational vacation landscape elements, including rough stones, Buddhism colors, sculptures, architectural pieces, totem columns, Tai style featured pavilion, blue waters and subtropical plants, all fully reveal the landscape in subtropical areas.

PART C　COMMERCIAL PROJECT
商业项目

11 南宁万达茂售楼处景观亭
12 南宁万达茂售楼处水景及露台
13 南宁万达茂售楼处水景
14 南宁万达茂售楼处景观

12

13

14

2014 WANDA COMMERCIAL PLANNING
万达商业规划——销售类物业

SALES OFFICE OF CHANGSHU WANDA PLAZA
常熟万达广场售楼处

开放时间 2014 / 09 / 20　**OPENED ON** SEPTEMBER 20 / 2014
项目位置 江苏 / 常熟　**LOCATION** CHANGSHU / JIANGSU PROVINCE
占地面积 1714 平方米　**LAND AREA** 1,714m²
建筑面积 2849 平方米　**FLOOR AREA** 2,849m²

BUILDING OF SALES OFFICE
售楼处建筑

立面设计采用新古典主义风格，通过玻璃幕墙、铝板等现代材料演绎新古典主义建筑内在神韵。售楼处主入口采用黑镜面钢材与玻璃结合，极具标志性。两侧建筑采用传统古典元素，厚重大气，具有仪式感，更加突出主入口的震撼效果。

The facade design adopts neoclassicism style. The inner charm of neoclassical building is interpreted through the application of modern materials including glass curtain wall and aluminum plate. The iconic main entrance of the sales office combines the black-mirror steel and glass. The buildings on both sides, which adopt traditional classical elements, look dignified and grand and present a ritual sense, giving greater prominence to the shocking effects of the main entrance.

01

PART C | COMMERCIAL PROJECT 商业项目

172 / 173

01 常熟万达广场售楼处总平面图
02 常熟万达广场售楼处外立面

04

05

INTERIOR OF SALES OFFICE
售楼处内装

室内设计采用现代欧式风格，严格控制空间的秩序感与中心的对称性，充分利用大堂两层挑高的空间优势。以活跃、明亮的对比色彩为基底，采用天然大理石、玫瑰金镜面及细腻的不锈钢线条，营造热烈的销售氛围。

Adopting modern simple European style, the interior design applies strict control on the spatial order impression and the central symmetry, making the best of the spatial advantage of the high and open lobby. Taking vibrant and bright contrasting colors as base, an enthusiastic sales atmosphere is created with the adoption of natural marble, rose gold mirror plane and delicate stainless steel lines.

06 常熟万达广场售楼处景观字体雕塑
07 常熟万达广场售楼处景观木平台
08 常熟万达广场售楼处景观绿化
09 常熟万达广场售楼处景观广场

LANDSCAPE OF SALES OFFICE
售楼处景观

景观设计风格与建筑风格相契合。通过开敞入口空间和具有导向性的铺装，起到吸引人流和导向作用。以木平台、绿化、小型广场为主的空间组合，在满足功能需求的基础上，运用艺术化、人性化的细节处理，营造舒适、浪漫的空间感受。

The design style of the landscape is consistent with that of the building. The open entrance space and the oriented pavement play a role of attracting people flow and guidance. On the basis of satisfying functional demands, the spatial organization, which is dominated by wood platforms, green space and small squares, creates a comfortable and romantic space perception through artistic and humanized details.

2014 WANDA COMMERCIAL PLANNING
万达商业规划——销售类物业

样板间 01

PROTOTYPE ROOMS OF DONGGUAN HOUJIE WANDA PLAZA
东莞厚街万达广场样板间

开放时间 2014 / 04 / 19	**OPENED ON** APRIL 19 / 2014
项目位置 广东 / 东莞	**LOCATION** DONGGUAN / GUANGDONG PROVINCE
建筑面积 229 平方米	**FLOOR AREA** 229m²

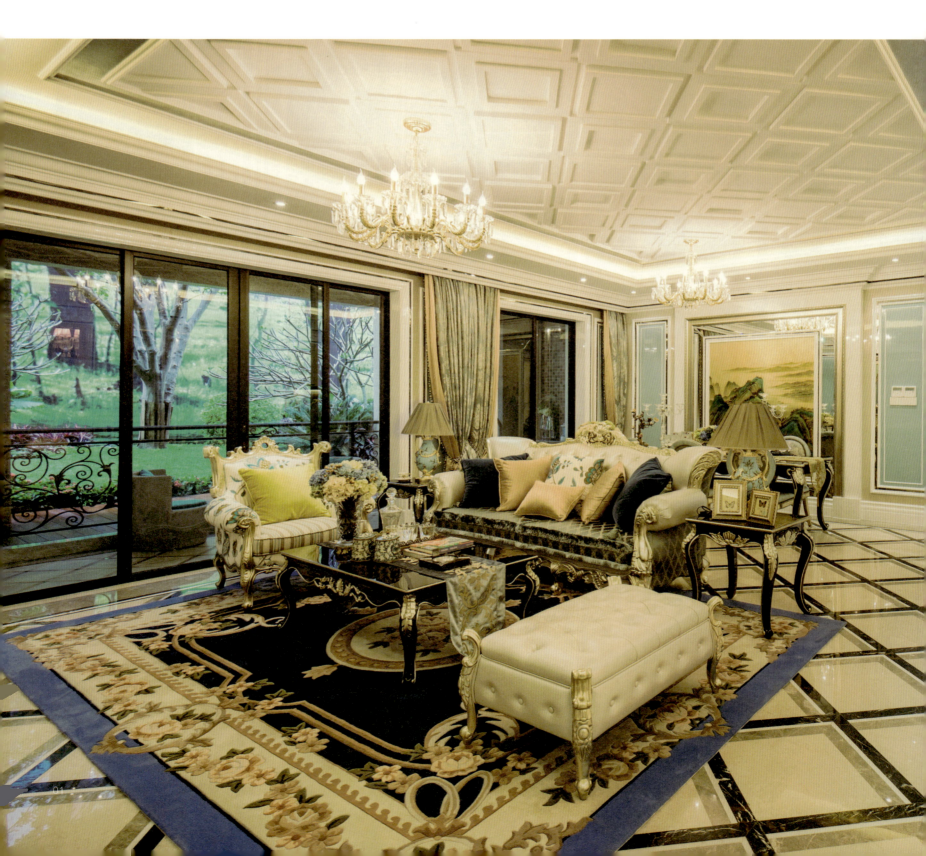

EUROPEAN-STYLE PROTOTYPE ROOM
欧式风格样板间

整体设计重视形式构图美，采用对称、变换、重复、组合多种手法，通过家具布局、装饰纹样，强化了形式构图的美感。

Adopting multiple techniques including symmetry, transformation, repetition and combination, the overall design highlights the beauty of form composition, which is further intensified through furniture layout and decorative patterns.

01 东莞厚街万达广场样板间 - 欧式风格样板间客厅
02 东莞厚街万达广场样板间 - 欧式风格样板间卧室
03 东莞厚街万达广场样板间 - 欧式风格样板间手绘效果图

04 东莞厚街万达广场样板间－现代风格样板间客厅
05 东莞厚街万达广场样板间－现代风格样板间户型图
06 东莞厚街万达广场样板间－现代风格样板间书房
07 东莞厚街万达广场样板间－现代风格样板间卧室

MODERN STYLE PROTOTYPE ROOM
现代风格样板间

以黑、白、卡布奇诺为基调，营造低调奢华的现代简约氛围。通过精细的工艺做法达成简约大气的展示效果。

A modern simple atmosphere of modest luxury is established with the base colors of black, white and cappuccino, and a simple yet grand exhibition effect is achieved through elaborate techniques and processes.

PART C COMMERCIAL PROJECT
商业项目

2014 WANDA COMMERCIAL PLANNING
万达商业规划——销售类物业

样板间 02

PROTOTYPE ROOMS OF ZHENGZHOU JINSHUI WANDA PLAZA
郑州金水万达广场样板间

开放时间 2014 / 05 / 21	OPENED ON MAY 21 / 2014
项目位置 河南 / 郑州	LOCATION ZHENGZHOU / HENAN PROVINCE
建筑面积 430 平方米	FLOOR AREA 430m²

01 郑州金水万达广场样板间-欧式古典风格样板间卧室
02 郑州金水万达广场样板间-欧式古典风格样板间客厅

CLASSIC EUROPEAN-STYLE PROTOTYPE ROOM
欧式古典风格样板间

以门厅为中轴线,划分客厅和餐厅区域。设计摒弃了繁复的描金装饰手法,保留顶棚优雅的灯槽线脚,墙面的精致造型和欧式的壁纸,让空间更具温馨浪漫气息。

The living room area and dining room area are divided by the hallway as a central axis. Abandoning the complicated gold outline decoration technique and keeping the elegant overlaid lines of the ceiling, the delicate modeling of the wall and the European-Style wall paper render the space a warm and romantic atmosphere.

03 郑州金水万达广场样板间-法式经典风格样板间书房
04 郑州金水万达广场样板间-法式经典风格样板间卧室
05 郑州金水万达广场样板间-法式经典风格样板间卧室
06 郑州金水万达广场样板间-法式经典风格样板间门厅

CLASSIC FRENCH-STYLE PROTOTYPE ROOM
法式经典风格样板间

镶嵌整块晶莹玉石的法式端景墙,像徽章一样以石材拼花铺就的地面,层次分明的顶棚线脚,温馨典雅的窗帘布艺,精挑细选的装饰摆件——这些细节都刻画了法式古典的富丽和精致。

The details of French-style feature wall inlaid with a whole piece of sparking and crystal-clear jadestone, the floor paved with stones in mosaic pattern like a badge, the well-arranged lines of the ceiling, the cozy and elegant curtain and the selective decoration ornaments together help to depict a magnificent and delicate Classic French Style.

2014 WANDA COMMERCIAL PLANNING
万达商业规划——销售类物业

PROTOTYPE ROOMS OF SHENYANG OLYMPIC WANDA PLAZA
沈阳奥体万达广场样板间

开放时间 2014 / 05 / 20　　**OPENED ON** MAY 20 / 2014
项目位置 辽宁 / 沈阳　　**LOCATION** SHENYANG / LIAONING PROVINCE
建筑面积 298 平方米　　**FLOOR AREA** 298m²

01

02

CLASSIC EUROPEAN-STYLE PROTOTYPE ROOM
欧式古典风格样板间

餐厅与客厅连通，宽敞明亮，动线流畅，空间统一性强。墙面运用了经典的欧式线脚及造型，体现气派和风范。顶棚木雕工艺细腻，融入洛可可装饰元素。餐厅木饰面和具有立体感的墙纸，加上正中间镶嵌的欧式油画，配合华丽的水晶灯、奢华的餐桌椅、餐具，营造出宫廷仪式感。

With dining room and the living room connected, spacious and bright space, smooth circulation and high space uniformity are achieved. Adopting classic European-style lines and shape, the wall is grand and magnificent. The ceiling adopts delicate wood carving technique with integration of Rococo decorative element. A ritual sense of royal court is created by the wood facing, the stereoscopic wall paper, along with the European-style oil painting inlaid in the middle and the elegant chandelier and luxurious dining table and chairs and tableware of the dining room.

03

01 沈阳奥体万达广场样板间-欧式古典风格样板间走廊
02 沈阳奥体万达广场样板间-欧式古典风格样板间卧室
03 沈阳奥体万达广场样板间-欧式古典风格样板间户型图
04 沈阳奥体万达广场样板间-欧式古典风格样板间客厅

COMMERCIAL PROJECT
商业项目

05 沈阳奥体万达广场样板间-欧式古典风格样板间客厅
06 沈阳奥体万达广场样板间-欧式古典风格样板间卧室
07 沈阳奥体万达广场样板间-欧式古典风格样板间卧室

2014 WANDA COMMERCIAL PLANNING
万达商业规划——销售类物业

PROTOTYPE ROOMS OF LONGYAN WANDA PLAZA
龙岩万达广场样板间

开放时间 2014 / 08 / 30　　**OPENED ON** AUGUST 30 / 2014
项目位置 福建 / 龙岩　　**LOCATION** LONGYAN / FUJIAN PROVINCE
建筑面积 825 平方米　　**FLOOR AREA** 825m²

01 龙岩万达广场样板间－创意公司样板间洽谈区
02 龙岩万达广场样板间－创意公司样板间会议室
03 龙岩万达广场样板间－创意公司样板间办公区

CREATIVE COMPANY PROTOTYPE ROOM
创意公司样板间

样板间设计为高端创意类公司办公空间。浅色高光影木运用点、线、面设计奠定了空间基调，功能布局与空间动线充分考虑了创意类公司的使用特点，营造了具有国际视野、现代感十足的高端办公室形象。

The prototype room is designed as an office space for high-end creative companies. The spatial keynote is established through point, line and surface design of light-color specular wood, and the functional layout and spatial circulation give full consideration to the usage characteristics of creative companies, which as a whole create an image of a high-end international and modern office.

HEADQUARTERS OFFICE PROTOTYPE ROOM
企业总部办公室样板间

样板间设计主题为企业总部办公,以展示企业形象和接待为主。通过石材、皮革、木饰面等高端材料的有机结合,营造稳重、大气的企业总部形象。

The prototype room is designed as headquarters office mainly used for the exhibition of enterprise image and reception. Through an organic combination of quality materials including stone, leather and wood facing, a dignified and grand image of headquarters is created.

PART C COMMERCIAL PROJECT
商业项目

04 龙岩万达广场样板间-企业总部办公室样板间会客区
05 龙岩万达广场样板间-企业总部办公室样板间会议室
06 龙岩万达广场样板间-企业总部办公室样板间卧室

2014 WANDA COMMERCIAL PLANNING
万达商业规划——销售类物业

样板间 05

PROTOTYPE ROOMS OF SHUDU WANDA PLAZA
蜀都万达广场样板间

开放时间 2014 / 08 / 16	**OPENED ON** AUGUST 16 / 2014
项目位置 四川 / 成都	**LOCATION** CHENGDU / SICHUAN PROVINCE
建筑面积 298 平方米	**FLOOR AREA** 298m²

HONG KONG STYLE PROTOTYPE ROOM
港式风格样板间

针对具有一定经济实力的中青年客户。功能空间、装饰风格、家具饰品的设计，注重满足客户享受生活、追求社会地位和认同感的需求。

This type targets at young and middle-aged customers with a fair economic strength. The design of function space, decorative style and furniture and decorations stresses on satisfying customer's demands of enjoying life and pursuing social status and sense of identification.

01 蜀都万达广场样板间－港式风格样板间客厅
02 蜀都万达广场样板间－港式风格样板间餐厅
03 蜀都万达广场样板间－港式风格样板间书房
04 蜀都万达广场样板间－港式风格样板间卧室

SIMPLE EUROPEAN STYLE PROTOTYPE ROOM
简欧风格样板间

该样板间设计风格定位为现代维多利亚风格,旨在满足城市新中产阶级改变居住环境和室内装饰样式的追求。

The design of this prototype room is positioned at modern Victorian style, aiming to satisfy urban new middle class's pursuit of upgraded living environment and interior decoration style.

MODERN STYLE PROTOTYPE ROOM
现代风格样板间

现代时尚的风格定位，为寻求都市潮流生活的青年一代营造专属空间。设计力求满足客户喜欢时尚、潮流，强调个性与独特的特点。

The style is positioned as modern and fashionable to create a dedicated space for young generation who seeks urban stylish life. The design strives to satisfy their appreciation of fashion and trend and their emphases on individuality and uniqueness.

05 蜀都万达广场样板间-简欧风格样板间客厅
06 蜀都万达广场样板间-简欧风格样板间卧室
07 蜀都万达广场样板间-现代风格样板间客厅
08 蜀都万达广场样板间-现代风格样板间书房

2014 WANDA COMMERCIAL PLANNING
万达商业规划——销售类物业

PROTOTYPE ROOM OF SHANGYU WANDA PLAZA
上虞万达广场样板间

开放时间 2014/08/31	**OPENED ON** AUGUST 31 / 2014
项目位置 浙江/绍兴	**LOCATION** SHAOXING / ZHEJIANG PROVINCE
建筑面积 130 平方米	**FLOOR AREA** 130m²

01

SIMPLE EUROPEAN-STYLE PROTOTYPE ROOM
简欧风格样板间

以深金为主色调，配以稳重的深啡色实木地板，点缀以轻快明亮的饰品、水墨格调的壁纸，让居室更加富有内涵。地面、顶棚简洁明朗，家具曲线柔美，让空间层次更加丰富。

With dark gold as the dominant color, matched with steady dark coffee solid wood floor and embellished with lively and bright ornaments and wallpaper of Chinese ink painting style, the living room is rich in connotations. The concise and clear floor and ceiling and the furniture with graceful curves render richer levels to the space.

01 上虞万达广场样板间-简欧风格样板间手绘效果图
02 上虞万达广场样板间-简欧风格样板间客厅

03 上虞万达广场样板间-简欧风格样板间书房
04 上虞万达广场样板间-简欧风格样板间卧室
05 上虞万达广场样板间-简欧风格样板间餐厅

2014 WANDA COMMERCIAL PLANNING
万达商业规划——销售类物业

样板间 07

PROTOTYPE ROOMS OF KUNSHAN WANDA PLAZA
昆山万达广场样板间

开放时间 2014 / 12 / 30 **OPENED ON** DECEMBER 30 / 2014
项目位置 江苏 / 昆山 **LOCATION** KUNSHAN / JIANGSU PROVINCE
建筑面积 255 平方米 **FLOOR AREA** 255m²

CHINESE-STYLE PROTOTYPE ROOM
中式风格样板间

设计融入了昆山传统艺术"昆曲"及其他中式元素，空间呈现出精致的江南韵味。以淡蓝色、米色、咖色系构建出温馨舒适的家居环境，散发浓郁的中式文化气息和历史沉淀感，色彩也达到和谐统一。整体空间划分和功能布局满足业主多元化的使用需求。

Integrating the traditional art of Kunshan – "Kunqu Opera" along with other Chinese elements into the design, the space presents an exquisite ambience of the South Yangtze River. A warm and comfortable home environment is built with the colors of light blue, beige and coffee to give off a rich atmosphere of traditional Chinese culture and history and to achieve a harmony and unity in color. The overall space division and function layout satisfy owners' diversified use demand.

01

02 03

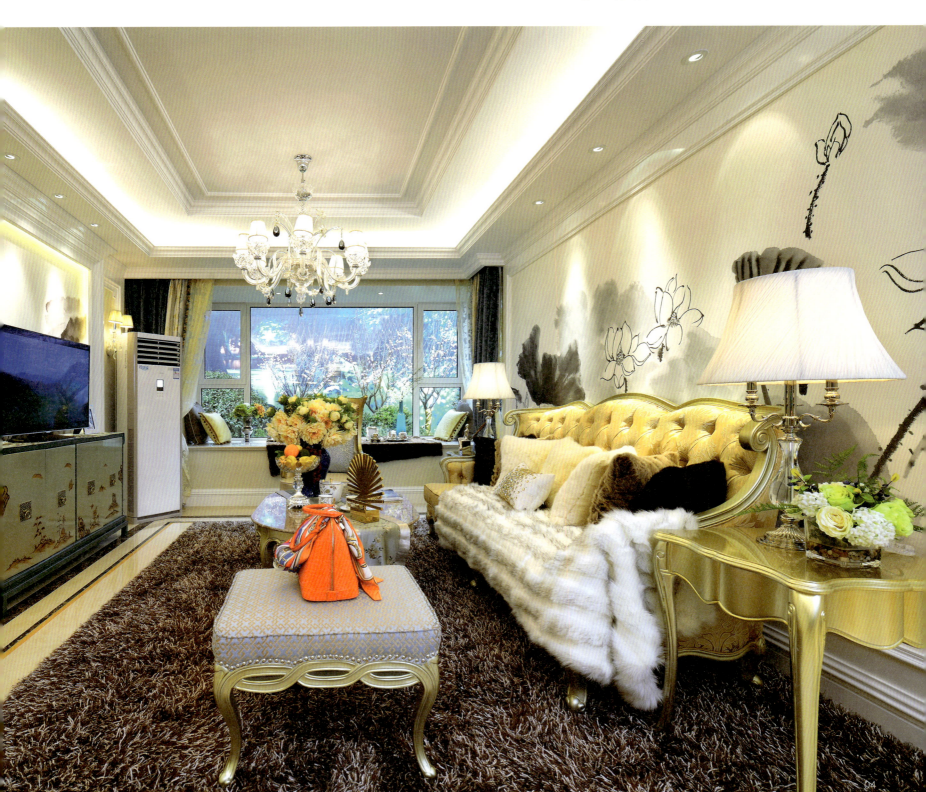

C1 昆山万达广场样板间-中式风格样板间户型图
02 昆山万达广场样板间-中式风格样板间卧室
03 昆山万达广场样板间-中式风格样板间书房
04 昆山万达广场样板间-中式风格样板间客厅

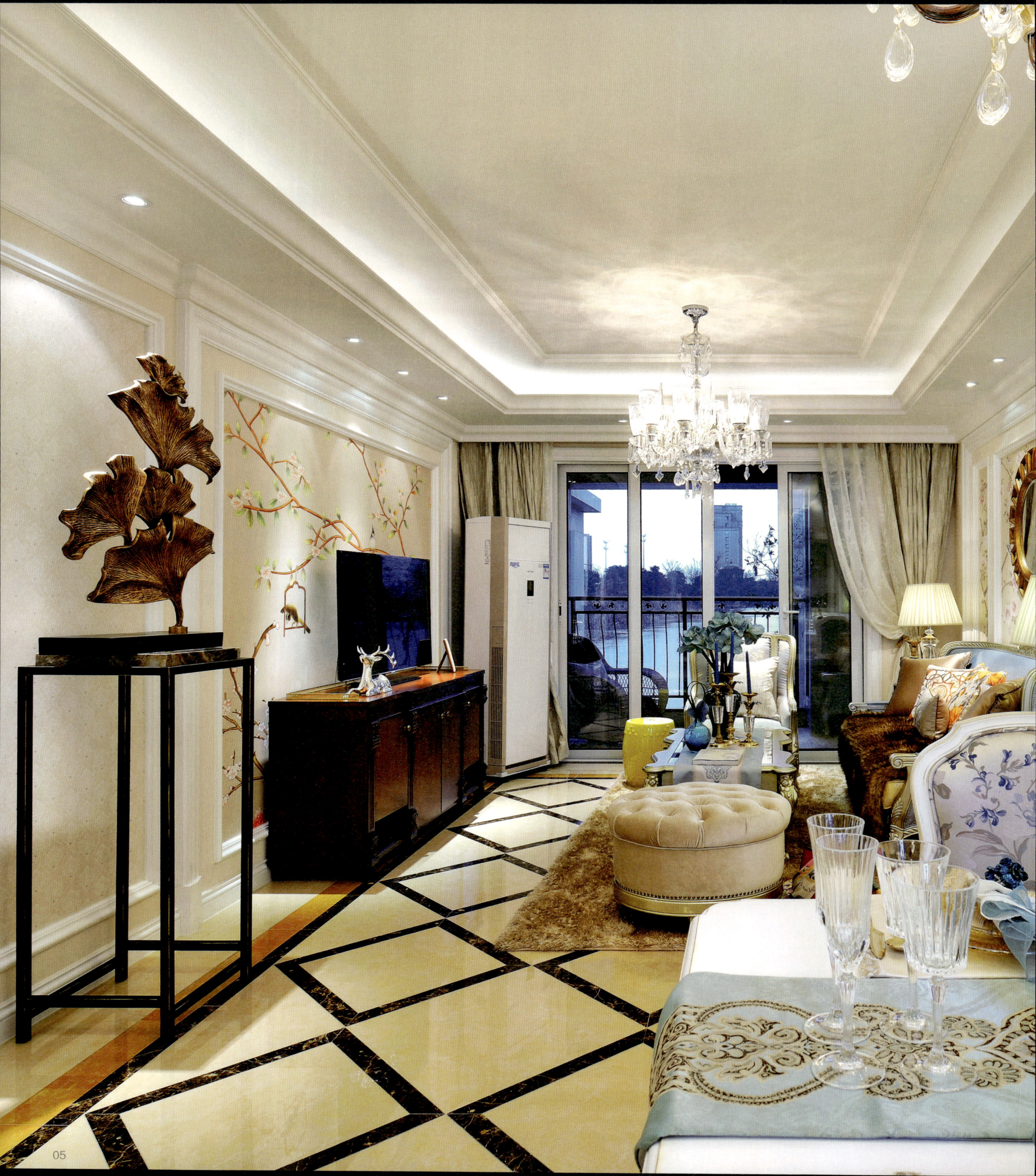

COMMERCIAL PROJECT
商业项目

SIMPLE EUROPEAN-STYLE PROTOTYPE ROOM
简欧风格样板间

具有田园气息的花鸟壁纸，粗旷自然的仿古砖、实木，休闲的家居摆设，让大自然的清新与惬意处处散放。简欧风格区别于传统欧式古典风格，它营造了轻松自在的生活氛围。

A fresh and cozy sense of the nature is felt through the pastoral wall paper of flowers and birds, the rough and natural archaized brick and solid wood and the casual household decoration. Different from traditional classic European style, simple European style intends to create comfortable and relaxed living atmosphere.

05 昆山万达广场样板间－简欧风格样板间客厅
06 昆山万达广场样板间－简欧风格样板间卧室
07 昆山万达广场样板间－简欧风格样板间卧室

2014 WANDA COMMERCIAL PLANNING
万达商业规划——销售类物业

PROTOTYPE ROOMS OF CHONGQING BA'NAN WANDA PLAZA
重庆巴南万达广场样板间

开放时间 2014 / 11 / 28	OPENED ON NOVEMBER 28 / 2014
项目位置 重庆 / 巴南	LOCATION BA'NAN / CHONGQING
建筑面积 110 平方米	FLOOR AREA 110m²

01 重庆巴南万达广场样板间-茶室样板间
02 重庆巴南万达广场样板间-公寓样板间户型图

TEAHOUSE PROTOTYPE ROOM
茶室样板间

茶室样板间为新中式风格，冷调的青砖、粗犷的麻石和温馨的木质、柔和的夹纱材质相碰撞，让整体空间丰富多变。经典的中式百草柜变身为展架，茶叶、茶具成为艺术品，博古架也华丽转型——这些匠心独具的设计，无一不体现着主人对艺术文化的追求与新的见解。

The teahouse prototype room adopts new Chinese style. With a conflicting utilization of cold-tone blue brick, rough granite, comfortable timber and soft mingled yarn material, the whole space is filled with diverse elements and changes. The classic Chinese-style cabinet for traditional Chinese herbs is served as a display rack; tea and tea sets are exhibited as works of art; the antique shelve undergoes a luxuriant transformation. These unique designs without exception reflect the owner's pursuit and new understanding of art and culture.

03 重庆巴南万达广场样板间-茶室样板间客厅
04 重庆巴南万达广场样板间-茶室样板间门厅
05 重庆巴南万达广场样板间-茶室样板间陈设
06 重庆巴南万达广场样板间-公寓样板间餐厅
07 重庆巴南万达广场样板间-公寓样板间卧室
08 重庆巴南万达广场样板间-公寓样板间客厅

APARTMENT PROTOTYPE ROOM
公寓样板间

公寓样板间为现代简约风格。主人的职业背景定位为珠宝设计师——热衷追求生活品质和艺术品位。以暖色为基调，在简单的设计中，增加细节与品质。门厅柜和全身镜为主人回家的"接风"、离家的"送行"，层板上展示的三人作品、工作桌上放置的手稿，均体现主人对工作的热情。

The apartment prototype room adopts modern simple style and defines the career background of the owner as a jewelry designer, who is enthusiastic about the pursuit for life quality and artistic taste. Taking warm colors as dominant colors, details and quality have been attached importance to in the simple design. The sideboard and full-length mirror stand in place to welcome the returned owner and see him off; the works of the owner exhibited on the shelf board and the manuscripts laid on the worktable both reflect the owner's passion for work.

2014 WANDA COMMERCIAL PLANNING
万达商业规划——销售类物业

PROTOTYPE ROOM OF DEZHOU WANDA PLAZA
德州万达广场样板间

开放时间 2014/05/31	**OPENED ON** MAY 31 / 2014
项目位置 山东/德州	**LOCATION** DEZHOU / SHANDONG PROVINCE
建筑面积 160平方米	**FLOOR AREA** 160m²

01 德州万达广场样板间-黑陶工作室样板间
02 德州万达广场样板间-黑陶工作室样板间工作台
03 德州万达广场样板间-黑陶工作室样板间陈设

BLACK POTTERY STUDIO PROTOTYPE ROOM
黑陶工作室样板间

样板间采用契合当地文化的"黑陶"为主题,打造了以展示、交流为主要功能的私人工作室。运用优雅创新的新中式风格,贯穿返璞归真的自然主义理念,在充分展示了功能空间的同时,也深刻地诠释了德州黑陶独有的文化内涵。

Adopting local-culture-oriented "black pottery" as its theme, the prototype room builds a private studio whose main functions are exhibition and communication. Adopting an elegant and creative new Chinese style and penetrating the naturalism philosophy that stresses the return to one's original nature, the space not only fully exhibits its function, but also profoundly interprets the exclusive cultural connotation of Dezhou black pottery.

2014 WANDA COMMERCIAL PLANNING
万达商业规划——销售类物业

样板间 10

PROTOTYPE ROOMS OF YINGKOU WANDA PLAZA
营口万达广场样板间

开放时间 2014 / 07 / 30　　**OPENED ON** JULY 30 / 2014
项目位置 辽宁 / 营口　　　　**LOCATION** YINGKOU / LIAONING PROVINCE
建筑面积 141平方米　　　　**FLOOR AREA** 141m²

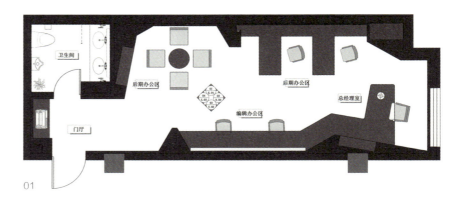

01

AUTOMOBILE ADVERTISING AGENCY PROTOTYPE ROOM
汽车广告公司样板间

内装设计选用"法拉利红"作为主色调，空间中也一再出现"F1赛道"的符号，通过强烈的色彩对比，使得小空间也显得活力十足。

Adopting "Ferrari Red" as the dominant color, the symbol of "F1 Racing Track" appear repeatedly in the interior space. Through strong color contrast, the small space appears to be dynamic and vibrant.

02

CLOTHING STUDIO PROTOTYPE ROOM
服装工作室样板间

服装工作室主题样板间是一个多变而灵动的空间，将服装制作中随手可取的布料、纽扣和丝线巧妙地转化为装饰材料，用缤纷的色彩语言雕琢空间，营造出高端、时尚、创意十足的私家服装工作室氛围。

The clothing studio prototype room is a changeable and dynamic space, which skillfully turns the cloth, buttons and silk threads at hand in cloth-making into decorative materials. Embellished with various colors, the space is built to be an elegant, fashionable and creative private clothing studio.

01 营口万达广场样板间-汽车广告公司样板间户型图
02 营口万达广场样板间-汽车广告公司样板间
03 营口万达广场样板间-服装工作室样板间工作台
04 营口万达广场样板间-服装工作室样板间

03

04

2014 WANDA COMMERCIAL PLANNING
万达商业规划——销售类物业

样板间 11

PROTOTYPE ROOM OF DONGYING WANDA PLAZA
东营万达广场样板间

开放时间 2014 / 04 / 23　　**OPENED ON** APRIL 30 / 2014
项目位置 山东 / 东营　　**LOCATION** DONGYING / SHANDONG PROVINCE
建筑面积 52 平方米　　**FLOOR AREA** 52m²

01 东营万达广场样板间-早教中心样板间
02 东营万达广场样板间-早教中心样板间
03 东营万达广场样板间-早教中心样板间

02

03

EARLY EDUCATION CENTER PROTOTYPE ROOM
早教中心样板间

样板间内装设计充分考虑了儿童的认知行为和兴趣取向，应用曲线布局和丰富的色彩，以《白雪公主》和《小熊维尼》的故事作为载体，以森林童话为空间主题，精心设计再现了一个充满童趣和易学易用的儿童空间。

The interior design of the prototype room gives full consideration to children's cognitive behavior and interest orientation. Adopting an curve layout and abundant colors and using the story of "Snow White" and "Winnie the Pooh" as carriers, a woods-fairy-tale-themed and easy-to-use Children's space filled with the fun of childhood is created through meticulous design.

2014 WANDA COMMERCIAL PLANNING
万达商业规划——销售类物业

实景示范区 **01**

DEMONSTRATION AREA OF DALIAN HIGH-TECH WANDA MANSION
大连高新万达公馆实景示范区

开放时间 2014 / 07 / 30	OPENED ON JULY 30 / 2014
项目位置 辽宁 / 大连	LOCATION DALIAN / LIAONING PROVINCE
占地面积 2 公顷	LAND AREA 2 HECTARES

PLANNING OF DEMONSTRATION AREA
示范区建筑规划

建筑采用19世纪欧洲新古典主义风格，强调空间造型的等级感、序列感与仪式感，强化轴线对景，讲求构图的形式美感，秩序严谨，体现出居住的身份感与房屋的价值感。建筑外形丰富而独特，形体厚重，线条鲜明，凹凸有致。

Adopting 19th Century European neoclassicism style, the building stresses the order perception, sequence sense and ritual sense of space, strengthens the symmetry axis opposite landscape layout and strives for the form beauty and rigorous order of composition to reflect the identity of the residents and the value of the building. The architectural appearance of the building is rich and unique with dignified shape and clear-cut and well-arranged lines.

01 大连高新万达公馆实景示范区总平面图
02 大连高新万达公馆实景示范区建筑外立面

2014 WANDA COMMERCIAL PLANNING
万达商业规划——销售类物业

03 大连高新万达公馆实景示范区鸟瞰图
04 大连高新万达公馆实景示范区鸟瞰图
05 大连高新万达公馆实景示范区建筑外立面

LANDSCAPE OF DEMONSTRATION AREA
示范区景观

景观采用法式古典主义风格，在空间格局上以凡尔赛宫为蓝本，体现端庄、大气的设计意向。整体以左右对称形式来表现，形成黄金中轴线，体现其宏伟壮观的景观效果。设计手法上注重细节和造型，给人以华美、精致、富丽堂皇的感觉。

The landscape adopts French-Style classicism style. The spatial framework models the Versailles to reflect a dignified and grand design intention. The overall landscape is in bilateral symmetry with a golden central axis, presenting a magnificent and splendid landscape effect. As for design skill, details and modeling are highlighted so as to create a sense of magnificence, delicacy and grandeur.

06 大连高新万达公馆实景示范区景观台阶
07 大连高新万达公馆实景示范区景观雕塑
08 大连高新万达公馆实景示范区景观亭

DEMONSTRATION AREA OF SHANGHAI JINSHAN WANDA PLAZA
上海金山万达广场实景示范区

开放时间 2014/08/29　　OPENED ON AUGUST 29 / 2014
项目位置 上海/金山　　LOCATION JINSHAN / SHANGHAI
占地面积 0.25 公顷　　LAND AREA 0.25 HECTARE

BUILDING OF DEMONSTRATION AREA
示范区建筑

立面采用纯正的Art Deco建筑风格，着眼于现代风格演化的立面效果，使得轮廓分明，竖向分隔挺拔内敛。建筑顶部采用顶级材料，彰显尊贵典雅的调性与恒久的价值感。建筑整体造型端庄稳重，基座、墙身、檐口屋顶，色彩统一中又略有变化，使之稳重而不沉重，造就了宁静祥和的居住区气氛。

Adopting authentic Art Deco architectural style and with a view to the facade effects evolved from modern style, the scheme facade has a clear outline and a straight and restrained vertical division. Top-level materials are applied for building top to demonstrate a distinguished and elegant atmosphere and a lasting sense of value. The overall modeling of the building is dignified and steady, and the color of the foundations, the walls, the cornices and the roofs are slightly changeable in a general unification, making it steady but not ponderous and creating a quiet and peaceful living environment.

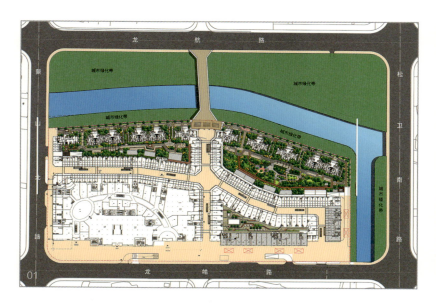

01 上海金山万达广场实景示范区总平面图
02 上海金山万达广场实景示范区住宅入口

LANDSCAPE OF DEMONSTRATION AREA
示范区景观

景观以"海上丝绸之路"为线索,从金山滨海文化中提取设计语言,围绕"海派时光"的文化主题进行整合设计,形成富有节奏的空间序列,营造丰富而灵动的商业空间。景观运用情境化小品与组团式绿化的融合,利用线性、几何感铺装的方向性指引,满足客户对高绿量、场景感、人性化空间的关注与需求。

Adopting "Maritime Silk Road" as a link, the landscape extracts design elements from Jinshan seaside culture and is designed comprehensively by centering on the cultural theme of "Shanghai Time" to form a rhythmic spatial series and create a rich and flexible commercial space. The landscape integrates situational landscape pieces with cluster greening and utilizes the directional guide of linear and geometric pavement to satisfy clients' concern and demand of high green quantity, scenes and humanized space.

03

04

03 上海金山万达广场实景示范区景观绿化
04 上海金山万达广场实景示范区景观花坛
05 上海金山万达广场实景示范区景观花坛
06 上海金山万达广场实景示范区景观小品

05

DEMONSTRATION AREA OF SHUDU WANDA PLAZA
蜀都万达广场实景示范区

开放时间 2014 / 10 / 18   **OPENED ON** OCTOBER 18 / 2014
项目位置 四川 / 成都   **LOCATION** CHENGDU / SICHUAN PROVINCE
占地面积 0.45 公顷   **LAND AREA** 0.45 HECTARE

BUILDING OF DEMONSTRATION AREA
示范区建筑

广场商业街以蜀都文化为线索，挖掘独有的地方文化特色和人文风情，结合当地的建筑材料以及构造做法——如小青瓦、坡屋面、木格窗、马头墙、青石板，并穿插欧式装饰元素，营造出现代与传统相结合、迎合主流消费的新型商业街。

Taking "Shudu Culture" as a link, the commercial street of the plaza digs the exclusive local cultural features and customs and integrates local building materials and construction schedule, like the Chinese-style blue tile, the pitched roof, the wood grid window, the wharf wall and the blue flagstone with insertion of European-style decorative elements, a new-type commercial street that combines modernity and tradition and caters to mainstream consumption is built.

02

01 蜀都万达广场实景示范区室外步行街建筑外立面
02 蜀都万达广场实景示范区总平面图

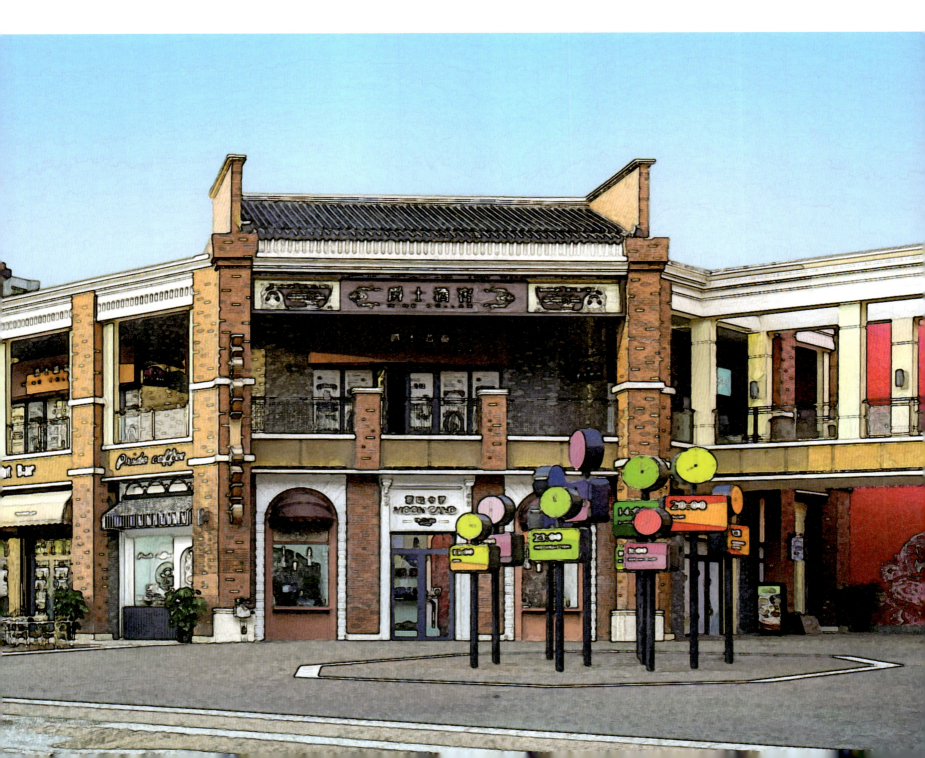

03

04

05

03 蜀都万达广场实景示范区室外步行街建筑外立面
04 蜀都万达广场实景示范区室外步行街建筑设计手绘效果图
05 蜀都万达广场实景示范区室外步行街夜景
06 蜀都万达广场实景示范区景观花园
07 蜀都万达广场实景示范区景观绿化
08 蜀都万达广场实景示范区水景

LANDSCAPE OF DEMONSTRATION AREA
示范区景观

广场实景示范区景观为现代田园特色自然式风格,布局自由、灵活,是天然山水的再现。本着"生态环保、用户导向、经济高效"的原则,结合动线合理设置特色瀑布、自然式跌水池、休闲大草坪、景观亲水平台、儿童活动场地等景观节点,将建筑、山水、花木合理地组合成一个生态综合体,在满足建造标准的同时亦具地方特色。种植层次错落有致,富有诗情画意。

The landscape of the plaza demonstration area adopts a natural style with modern pastoral features and free and flexible layout, which reproduces natural landscape. In adherence to the principle of eco-environment protection, client orientation and high economic efficiency and in combination with circulation, landscape nodes (including a feature waterfall, a natural style plunge pool, a leisure lawn, a landscape waterborne platform and a children activity space) are reasonably laid out to rationally integrate buildings, landscape and flowers and trees into an ecological complex. Besides, with well-organized and idyllic plantations, both construction cost standard and local features are satisfied.

2014 WANDA COMMERCIAL PLANNING
万达商业规划——销售类物业

实景示范区 04

DEMONSTRATION AREA OF SIPING WANDA PLAZA
四平万达广场实景示范区

开放时间 2014 / 07 / 20	OPENED ON JULY 20 / 2014
项目位置 吉林 / 四平	LOCATION SIPING / JILIN PROVINCE
占地面积 0.55 公顷	LAND AREA 0.55 HECTARE

LANDSCAPE OF DEMONSTRATION AREA
示范区景观

示范区景观主要分为入口景观区和中心景观区两部分。对称式景观树池分布于道路两侧，展现出入口景观区的仪式感和领域感。中心景观区设置了拾阶而上的中心草坪、精致大气的欧式喷泉。人们可以在草坪两侧的廊架下休憩，看到开阔的草坪、闻到花架上弥漫的花香，感受到潺潺流水带来的清新——这些细节，使示范区景观空间丰富、布局主次分明。

The demonstration area landscape is mainly divided into two parts – the entrance landscape area and the central landscape area. Symmetric landscape planters are distributed on both sides of the road, reflecting the ritual sense and territory sense of the entrance landscape area. A terraced central lawn and a delicate and grand European-style fountain are set at the central landscape area. People resting under the pergolas on either side of the lawn can enjoy the view of the spacious lawn, the pervasive fragrance of the flowers on the pergola and the feeling of freshness brought by the gurgling water. All these details make the demonstration area rich in space and clear in priority.

01 四平万达广场实景示范区总平面图
02 四平万达广场实景示范区景观喷泉
03 四平万达广场实景示范区景观花坛
04 四平万达广场实景示范区景观绿化
05 四平万达广场实景示范区景观设计手绘剖面图

EPILOGUE
后续

WANDA
COMMERCIAL
PLANNING
2014

DESIGN CONTROL OF PROPERTIES FOR SALE
销售物业设计管控

万达商业地产设计中心常务副总经理　林树郁
兼技术管理部总经理

万达商业地产设计中心承担着万达集团所有销售类物业及其配套的设计管控，全年管控项目134个，包含10个文旅项目和5个境外新项目，管控面积达1166.86万平方米。在集团领导的关心和设计中心同仁的努力下，设计中心圆满完成全年任务。本文介绍一下背后"助力"取得这些成绩的管控工具、管控要点和重要项目管控。

一、管控工具

管控工具分为集团管理制度、设计中心根据销售物业的管控工作需要编制的管控手册和工时管理三类。

1.万达集团管理制度

《万达集团管理制度》第二册为《商业地产管理制度》。该册第二章"项目设计管理"，从管理体系、设计管理权责界面、设计管理流程、设计变更管理、设计类业务发包管理和考核办法等六个方面对项目管理做了规定。第三册为《安全管理制度》，对项目和公司的安全管理做了规定。

2.设计中心设计管控手册

《销售物业设计管控手册》分为《流程及制度篇》、《建筑篇》、《精装篇》、《景观篇》、《结构篇》、《机电篇》和《绿建篇》七个分册，从方案前期、方案、方案深化、施工图、实施及考核六个阶段，按照"管理流程"、"管控标准"及"考核标准"三个维度，制定了覆盖销售物业建筑专业的设计管控文件。

3.设计中心品质管控手册

《销售物业品质手册》展示了项目操作过程中优劣做法及案例，明确了品质要求、工艺做法以及完成标准，可供相关管理人员参考。手册分为《建筑专业》、《精装专业（硬装）》、《精装专业（软装）》和《景观专业》四个分册。各册均包括"产品效果实例"及"验收表单"两部分。

4.工时管理理念

2014年8月，配合管控职能及人员编制调整，设计中心建立了量化工作的"工时统计"管理体系。该管理由设计中心每位成员提供日常工作事项、项目管控节点、各阶段、各事项用耗时量的完整统计，从而制

I. CONTROL TOOLS

Wanda Commercial Real Estate Design Center assumes the design control for all properties for sale and supporting facilities of Wanda Group, controlling 134 projects with area of 11.6686 million square meters for the whole year, wherein 10 cultural tourism projects and 5 new overseas projects are included. Under the care of Wanda Group leaders and endeavor of Design Center staff, Design Center has successfully completed the tasks throughout the year. This text intends to elaborate on control tools, control points and important projects control that are served as the "Booster" for achievements as such.

Control tools consist of Wanda Group Management System, Control Manual prepared as required by control work of properties for sale, and Man-Hour Management.

1. WANDA GROUP MANAGEMENT SYSTEM

In the second chapter entitled "Project Design Management" of *Commercial Real Estate Management System*, Volume B of *Wanda Group Management System*, rules are set out on project management from six aspects, respectively being the management system, design management responsibilities interface, design management process, design change management, design business contract management, and evaluation methods. In *Safety Management System*, Volume C of *Wanda Group Management System*, rules are made on safety management of project and company.

2. DESIGN CONTROL MANUAL OF DESIGN CENTER

Properties for Sale Design Control Manual encompasses seven books, including *Process and System, Architecture, Landscape, Structure, M&E* and *Green Building*. In accordance with the three dimensions of "Management Process", "Control Standard" and "Evaluation Standard", the manual formulates the design control document that covers architectural specialty of properties for sale in six phases, involving preliminary design, design, design details, construction drawing, implementation and evaluation.

3. QUALITY CONTROL MANUAL OF DESIGN CENTER

Properties for Sale Quality Control Manual delivers the pros and cons and cases of project operation, and specifies the quality requirements, technology practice and completion standard to be referred to by related management personnel. The Manual is divided into four books, including *Architectural, Fitting-Out (Hard Outfit), Fitting-Out (Soft Outfit)* and *Landscape*, each containing two parts of "Product Effect Case" and "Acceptance Form".

4. MAN-HOUR MANAGEMENT CONCEPT

In August 2014, to cooperate with its control function and staffing adjustment, the Design Center established "Man-Hour Statistic" management system to quantify work. Based on the global statistics for time cost of daily work items, project control nodes, each phase and each item provided by each member in Design Center, "Man-Hour Model" that applies standard work as per project quantity is made. At

工作分项	工时(h)	占比（%）
前期	3506	1.24
总图	31921	10.25
销售卖场	30202	10.72
方案	122193	36.46
深化方案	2145	0.76
施工图	7573	2.69
实施	27522	9.77
科研工作	9000	3.19
内部管理	63241	22.44
其他	6981	2.48
合计	311265	100

（表1）工时管理

定根据项目数套用标准工作的"工时模型"。同时，定期整理"工时模型"，并建立各部门的"工时数据库"，促进了岗位管理者的科学决策水平（表1）。

二、管控要点

1.设计前期

（1）规划风险管控

在项目正式启动前，项目公司应严格按照《规划风险管控要点表》要求，落实并填写各项规划风险管控要点。主要内容是：规划指标、建筑设计、市政配套、地块条件。

（2）设计前期条件调查

项目公司成立后，应立即按照《设计前期条件调查表》要求，落实场地资料、土地、规划、产品设计的要求，以及消防、人防的要求，符合市政、节能、交通、结构安全等各项前期条件。

2.卖场管控

（1）卖场选址

项目公司必须提前熟悉项目情况、完成销售卖场定位与选址；如果在地块外建设需上报商业地产总裁批准。

（2）方案及深化

项目公司应充分利用现有设计模块。如项目有特殊需求，应基于成型的设计模块进行优化、创新，并报集团批准。

（3）设计封样

销售卖场实施周期短，如施工封样在计划节点内无法满足设计要求，应提前与集团做好沟通，并将施工封样工作前置，确保在计划周期内，严格按照设计封样效果实施。

（4）第三方审查

销售卖场施工图需要完整地履行第三方审查程序。

the same time, through regularly organizing "Man-Hour Model" and establishing the "Man-Hour Database" of each department, the scientific decision level of the managers is facilitated (see Table 1).

II. CONTROL POINTS

1. PRELIMINARY DESIGN

(1) Planning risk control:
The project company shall, before the project officially launches, implement and fill all control points of planning risk in strict accordance with the requirements of *Planning Risk Control Points Form*, mainly including planning indexes, architectural design, municipal facilities and land conditions.

(2) Preliminary design condition survey:
After its establishment, the project company shall promptly carry out the requirements on site data, land, planning, product design, fire protection, civil air defense as required by *Preliminary Design Condition Survey Form*, satisfying initial conditions concerning municipal, energy efficiency, transportation, structural safety, etc.

2. SALES PLACE CONTROL

(1) Site selection of sale place:
The project company must be acquainted with the project situation ahead, and complete sales sale place positioning and site selection; Off-site construction shall be reported to President of Commercial Real Estate for approval.

(2) Design and detail:
The project company shall make full use of existing design modules and apply optimization and innovation based on established design modules and submit to Wanda Group for approval in case of any project with specific requirement.

(3) Design-sealed sample:
Given the short implementation cycle of sales store, if construction-sealed sample fails to satisfy design requirement within planned nodes, it shall be negotiated with Wanda Group in advance and construction-sealed sample work shall be advanced. In this manner, it is guaranteed to implement in rigorous compliance with design-sealed sample effect within the planning cycle.

(4) Third Party Review:
Sales store construction drawing shall be fully subject to the Third Party Review Procedure, which focuses on appropriate treatment for intersecting technical measures of fitting-out and civil engineering structure, plumbing and M&E, floor height space.

(5) Construction drawing management:
Given the short review cycle of sales store construction drawing, the project company shall prepare in advance to provide construction drawing in an earliest manner. General contractor shall realize coordination with project company ahead to prepare for drawing review.

(6) Store quality:
The Design Center is in charge of the opening acceptance for sales store; sales office must pass acceptance check of Safety Supervision Department of Wanda prior to its opening. General contractor shall visit and learn the outstanding project cases of Wanda in advance to get informed of quality standard and guarantee quality of sales store.

(7) Store evaluation:
The Design Center shall make real-time evaluation scoring for implementation effect of sales store and notice at monthly meeting.

侧重点在精装专业与土建结构、水暖机电、层高空间等的相互技术交叉措施是否妥善处理。

（5）施工图管理
销售卖场施工图纸的审核周期较短，项目公司应提前做好准备，尽早提供施工图纸。总承包商应提前与项目公司对接，做好审图的准备工作。

（6）卖场品质
设计中心负责销售卖场的开放验收，售楼处也必须通过万达安监部的检查验收后方可开放。总承包商应提前对万达优秀项目案例进行参观学习，了解品质标准，确保销售卖场品质。

（7）卖场考核
设计中心对销售卖场的实施效果即时予以考核评分，月度会通报。

3.产品设计
（1）招投标管理
自2014年开始，设计中心对设计方案的招投标工作，尤其是对"一类标"实行了全面管控，编制了各项工作的管理办法，并于当年开始执行项目网上招标。

（2）中间成果论证会
设计中心编制《方案中间成果论证会审查要点》、《方案中间成果论证会会签单》。最为重要的事项包括：前期资料完整正确、符合签批总图指标、符合营销产品定位、满足当地政府要求、满足继续深化条件。

（3）方案成果交底会
《方案成果交底会审查要点》、《方案成果交底会会签单》是双方对方案成果的共同确认，由设计中心编制。

（4）施工图第三方审查
销售物业施工图由第三方审查，为此设计中心编制了《销售物业施工图第三方审查工作管理办法》、《销售物业施工图第三方审查要点》。

（5）施工图管理
对于实行"工程总包"的项目的施工图管理，要求施工总承包单位认真审图，并按照《施工图审查要点表》、《施工图评审会议记录表》逐项审查、填写，根据施工图纸设计管理流程，需按照审核图纸提交、审核意见提交及会审、图纸正式移交三个阶段进行。

3. PRODUCT DESIGN
(1) Tendering management:
Since 2014, the Design Center has applied comprehensive control on tendering of design plans, especially on "Class I Tender", prepared measures for administration of various works, and started online tendering for project in the same year.

(2) Intermediate achievement feasibility study meeting:
The Design Center has compiled *Review Points of Design Intermediate Achievement Feasibility Study Meeting* and *Validation and Signing Form of Design Intermediate Achievement Feasibility Study Meeting*. The most important matters include the integrity and accuracy of initial data, compliance with signed and approved master plan index and marketing product positioning, satisfaction with requirements of local government and continued deepening conditions.

(3) Design achievement clarification meeting:
Review Points of Design Achievement Clarification Meeting and *Validation and Signing Form of Design Achievement Clarification Meeting* mark the joint confirmation on design achievements of both parties and are prepared by the Design Center.

(4) Third party review on construction drawing:
In response to the third party review on construction drawing of properties for sale, the Design Center has prepared *Measures for Administration of Third Party Review on Construction Drawing of Properties for Sale* and *Review Points of Third Party Review on Construction Drawing of Properties for Sale*.

(5) Construction drawing management:
In terms of construction drawing management of general contracted projects, the construction main contractor is required to carefully review construction drawings, review and fill out as per *Construction Drawing Review Points Table* and *Record Table of Construction Drawing Review Meeting*, and in accordance with construction drawing design management process, carry out the work in three phases, including submission of review drawings, submission of review comments and joint review, and official handover of drawings.

4. SEALED SAMPLE MATERIALS
(1) Tendering-sealed sample:
Bidding plans for effect design must be enclosed with sealed sample materials during bidding to facilitate effect and cost control.

(2) Plan-sealed sample:
While handing over plans to the project company, the Design Center shall hand over design-sealed sample materials, which shall be provided as per schedule module completion standard.

(3) Construction-sealed sample:
The project company shall keep design-sealed sample intact at construction site for the convenience of check with construction-sealed sample.

(4) Mock-ups:
Mock-ups shall be constructed by the main contractor as per schedule module completion standard at site and upon completion, subject to acceptance and signed confirmation by the project company.

4.材料封样

（1）投标封样
效果类设计投标方案投标时必须附有材料封样，以便控制效果和成本。

（2）方案封样
设计中心向项目公司移交方案的同时移交设计封样。设计封样应按计划模块的完成标准提供。

（3）施工封样
项目公司应将设计封样在施工现场完好保存，以便于与施工封样进行核对。

（4）样板段
总包应在现场按照计划模块样板段完成标准进行样板段施工，并由项目公司在样板段完成后验收并签字确认。

5.成果移交
按时有序地完成规划院移交、设计中心移交、项目公司移交、总承包商移交等各项成果移交工作。

6.现场管控

（1）现场品质管控
确保现场严格按图纸及材料封样施工。对于未按图施工部分，项目公司有权要求总包按照原设计要求进行整改（含拆除）。

（2）现场安全管控
设计中心重点检查初设及安全评审会、施工图及安全评审会的设计安全类问题。做好现场检查中发现的设计安全问题与不按图施工问题的收集、整理、检查、销项。

7.设计变更
设计变更有明确流程要求。特别强调的是：应先申请再变更。变更申请前需确认：是否涉及面积指标变化、是否选用集团品牌库材料、是否与交房标准一致。

三、重要项目管控

大型文旅项目的操作极其复杂，设计中心采取的几项特殊措施，有效地保障了大型文旅项目高品质快速推进。

1.编制展示中心标准化设计模块手册
设计中心组织编写了标准化设计模块手册，以实现过程要点管控的标准化、流程化、制度化的目的。手册分为《方案册》和《技术册》两部分：

（1）《方案册》——包含展示中心方案总则、总图、

5. ACHIEVEMENTS HANDOVER
Achievements handover of Planning Institute, Design Center, Project Company and Main Contractor shall be carried out in order and on schedule.

6. ONSITE DESIGN CONTROL
(1) Onsite quality control:
Onsite construction as per drawings and sealed sample materials shall be guaranteed. As for the section failing to be constructed as per drawings, the project company is entitled to require main contractor to rectify it as per original design, including demolition.

(2) Onsite safety control:
The Design Center concentrates on checking design safety in preliminary design and safety review meeting, construction drawing and safety review meeting, accomplishing the collection, sorting, inspection and item list for design safety matters and construction incompliant with drawings spotted during onsite check.

7. DESIGN CHANGES
There is specific process requirement for design changes. It is worth noting that application is required prior to design change. And before application, confirm that whether the design change involves area index variation, selects materials included in Wanda's brand list, and corresponds with deliver standard.

III. IMPORTANT PROJECTS CONTROL

The Design Center takes several specific measures against extremely complex large-scale cultural tourism projects, effectively safeguarding the rapid advancement of projects as such with high quality.

1. PREPARATION OF STANDARD DESIGN MODULE MANUAL FOR EXHIBITION CENTER
The Design Center has prepared the standard design module manual to realize the standardized, processed and systematic process points control. The manual is composed of *Plan Manual* and *Technical Manual*.

(1) the *Plan Manual* includes standard modules for plan general, master plan, architectural design, exhibition design (smart building model; 3D experience hall; LED screen) of exhibition center, realizing standardized plan design points.

(2) the *Technical Manual* contains design control points of each specialty, construction drawing design standard, special review control process and template.

2. CONCEPTUAL PLANNING
At conceptual planning phase, the plan is required to give comprehensive consideration to full use of natural conditions and human resources at master plan phase, such as road network planning, earthwork balance, utilization of groundcover plants, preservation of famous and ancient trees, value exploration of relics sites.

3. PREPARATION OF LANDSCAPE DESIGN GUIDELINE
Landscape design guideline is prepared to specify project analysis, theme extraction, theme insertion, theme expression, etc.

4. ELEVATION PLANNING
Elevation planning is an integral part of cultural tourism project design. Each key project under the control of the Design Center, prior to land elevation design, shall undergo overall elevation planning.

建筑设计、展示设计（智能沙盘、3D体验厅、LED显示屏）的标准模块，实现方案设计要点标准化。

（2）《技术册》——包含各专业设计管控要点、施工图设计标准、专项评审管控流程及模板。

2. 概念规划

概念规划阶段，要求方案综合考虑总图阶段的路网规划、土方平衡、地被植物的利用、名木古树的保护、遗迹景点的价值挖掘等自然条件、人文资源的充分利用。

3. 编制景观设计导则

编制景观设计导则，对项目分析、主题提炼、主题嵌入、主题表现等做出规定。

4. 立面策划

立面策划是文旅项目产品设计的一项重要内容。设计中心管控的每一个重点项目，在开始地块立面设计前，都要先进行整体的立面策划。

（1）形象营造

以哈尔滨万达城为例，项目通过采用俄式风格立面元素及设计手法，在项目整体上体现具有俄式风情的单体和街区形象，作为整个策划及其落地实施的基础（图1）。

（2）主题营造

方案首先根据不同的街区主题，选取与主题相适合的典型风格建筑，然后对所选定的典型建筑进行立面造型元素的提取和归纳，在分类归纳的基础上进行具体的方案设计，最终形成意向立面和策划方案（图2）。

(1) Image building:
Through the adoption of Russian style elevation elements and design methods, the project presents the Russian style individual building and block image as a whole, making it the base for the whole planning and implementation(Fig.1).

(2) Theme building:
The design plan firstly selects typical style building suitable for diverse block themes, then extracts and summarizes elevation modeling elements from the selected typical building. Through specific plan design based on sorted generalization, intentional elevation and planning scheme are eventually formed(Fig.2).

(3) Atmosphere building:
Harbin Wanda Cultural Tourism City, for example, by introducing soft outfit elements of Russian style culture and art such as gardens, sculptures, murals, displays, formats, and even performance art, makes Russian style stand out and accomplished 3D construction and exhibition of Russian style. The cultural tourism feature of cultural tourism projects is thus highlighted(Fig.3).

5. SOFT OUTFIT DESIGN GUIDELINE

Soft outfit is mainly controlled by the following means.
(1) Standardizing the soft outfit design standard to prevent effect and cost getting out of control during soft outfit implementation.

(2) Establishing soft outfit plan design standard template to control the overall quality of soft outfit plan.

(3) Forming a unified standard through quantitative form, quantity control of furniture and accessories listed in soft outfit plan, helping to control components and missed items, improve the efficiency of management and avoid "unwanted goods" during implementation.

(4) Formulating soft outfit furniture selection standard to control effect and quality of furniture.

（图1）哈尔滨万达城立面策划定稿主题

  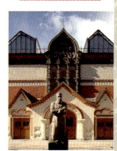

圣瓦西里教堂　　套娃展览馆　　普希金造型艺术馆　　特列季亚科夫画廊

（图2）街区主题营造

（3）氛围营造

通过引入俄式风情文化、艺术等软件元素，如园林、雕塑、壁画、美陈、业态，甚至是行为艺术等，烘托俄式风情文化，形成对俄罗斯风情的立体营造和展示，凸显文旅项目的文化旅游特质（图3）。

5.软装设计导则

软装方面主要通过如下手段进行管控：

（1）规范软装方案设计标准，避免软装实施过程效果及成本失控；

（2）制定软装方案设计标准模板，通过模版控制软装方案整体品质；

（3）通过量化表单、控制软装方案中家具、饰品数量，形成统一标准，便于控制部品漏项，提高管理效率，避免实施过程"货不对版"；

（4）制定软装家具选型标准，通过选型标准控制家具效果品质。

（图3）哈尔滨万达城立面策划定稿主题

IV. CONCLUSION

In 2014, the Design Center adjusts from internally to the development needs of the society and Wanda Group. We are determined to face new pattern, establish new model, form new thinking, harvest new achievement, and adapt to new normal.

四、结语

2014年，设计中心通过内部调整以适应社会和集团的发展需要。我们要面对新格局，建立新模式，形成新思维，取得新成果，适应新常态！

NEW TASKS OF PROJECT CONTROL OF CULTURE AND TOURISM PROPERTIES FOR SALE
文旅销售物业项目管控的新课题

万达商业地产设计中心副总经理兼文旅设计部总经理 门瑞冰

万达文旅项目始于2012年长白山万达国际旅游城，如今仅过了短短的三年，文旅万达城"在施"项目从最初的一个增加到了十个，总用地面积达1021公顷，总建筑规模达3426万平方米。在商业地产领域，这样的增长速度，全世界可能只有中国、全中国可能只有万达能够实现。当然，这样的飞速增长，对于万达来讲也是一份足以自豪的成绩。

文旅项目这样迅猛的增速、这么巨大的规模，也给项目的设计管控带来了巨大的压力。尤其是管控面积占到文旅项目的60%~70%、开发周期平均长达8~10年的文旅销售物业，其设计管控的难度、强度，在全国地产行业的设计管理系统中，都是名列前茅的。

回顾这一年的设计管控工作，文旅设计部在方案效果、品质监管、现场管控、计划管理等方面都做了大量的工作，不断提升设计管理技巧与力度，才能够取得上述成绩。具体来讲有以下几点。

1. 提高方案审核标准，加强方案审核力度，严格方案审核流程

为了提高文旅销售物业最终实施的效果及品质，设计中心文旅设计部首先在管控源头——方案效果上加强管控力度。

首先，提高对方案效果、品质的要求。无论是建筑、内装还是景观，哪怕只是一个局部方案，经常都要经过多轮优化调整、对比分析，直到达到理想的效果才允许进行下一步工作。

其次，加强方案审核的力度。要求方案无论大小都要经过审核才能移交。对于材料封样、构造做法的审核也更加严格，例如主要材料、对效果实现有关键影响的材料、关键节点构造做法，必须在现场制作实体样板段进行效果确认。例如广州万达城展示中心项目，就在施工初期在现场制作了1:1的幕墙实体样板段，对幕墙玻璃材料、夜景照明灯具安装方案及照明效果进行了验证，保证了展示中心展示效果的完美实现。

第三，文旅设计部在方案审核的流程上也加强了管理，要求报审方案必须经过专业副总、部门总经理两

Wanda's cultural tourism project can be dated back to the Changbai Mountain International Tourism Resort in 2012, since which the projects under construction of Wanda Cultural Tourism City (hereinafter referred to as Wanda City) has increased from the one to ten in only three years, with a total land area of 1021 hectares and total building scale of 34.26 million square meters. Looking around the commercial real estate industry, only China all over the world, only Wanda all over China are prone to achieving such rapid growth, which is indeed an achievement that Wanda should take pride in.

Obviously, the quickly booming and oversized cultural tourism project has also brought the project design control great pressure. Especially, the design covers 60%~70% of the cultural tourism projects with average development cycle being as long as 8 to 10 years, thus the difficulty and intensity in its design control top the design management systems in the national real estate industry.

Looking back to our design control works in the past year, the Cultural Tourism Design Department has invested a lot in scheme effect, quality supervision, onsite control, plan management, and strives to address the issues above through continuously enhancing design management skill and power. The key points are summarized as below.

1. APPLYING ENHANCED SCHEME APPROVAL STANDARD, REINFORCED SCHEME APPROVAL FORCE AND RIGOROUS APPROVAL PROCESS

To promote the ultimate implementation effect and quality of cultural tourism properties for sale, the Cultural Tourism Design Department of Design Center firstly vigorously strengthens its control on the origin, the scheme effect.

First, improve requirements on scheme effect and quality: buildings, interiors, landscapes or even partial scheme, before entering into the next phase, shall always go through several rounds of optimization, adjustment and comparative analysis until ideal effect is achieved.

Second, reinforce scheme approval force: all schemes shall be handed over upon approval irrespective of their size; the approval for material-sealed samples and construction practice shall be more rigorous. The main materials, materials largely determining the effect, construction practice of key nodes, for example, must be made on site with physical mock-ups for effect confirmation. Take Guangzhou Wanda Exhibition Center project for example. 1:1 curtain wall prototype section is constructed on site at the preliminary construction stage to verify glass materials of curtain wall, nightscape lighting fixtures scheme and lighting effect, finally guaranteeing the perfect implementation of the exhibition effect of the center.

Third, strengthen management on scheme approval process: the schemes submitted for approval must be reported to manager's drawing committee of Design Center upon

级审核后才能上报设计中心总经理审图会。所有上报营销会审核的方案，必须通过设计中心总经理审核后方能上会。通过对方案审核流程的严格要求，文旅设计部杜绝了集团领导对方案意见不同情况的发生，避免了后期实施方案整改的风险，确保了工程进度的按计划推进，降低了无效成本增加的风险。

2．通过立面策划、城市空间策划等工作，制定项目方案设计的指导性原则，实现项目品质、效果的统一、可控

设计中心文旅设计部从2014初开始进行文旅各个项目销售物业立面策划的工作。通过立面策划，对文旅各个项目的销售物业从文化的角度出发，对其空间主题、立面风格、街区形象进行统筹规划，使得后续各个地块的具体方案设计能够有统一的、可持续发展的指导性原则作为设计依据，从而确保项目整体效果、品质的协调统一。

立面策划工作及成果得到了集团领导的一致好评。根据领导对策划工作的具体指示，文旅设计部在立面策划的基础上，进一步开展了城市空间策划工作。作为立面策划的升级换代，城市空间策划能够从更宏观、更全面、更系统的层面对文旅销售物业的品质实现进行指导性管控。相信随着立面策划、城市空间策划的研究与实施，文旅销售物业的品质效果必将得到进一步提升。

3．加强现场管控，加强对项目公司设计部门的沟通、管理与培训，提升现场管控水平，确保方案效果与品质的实现

设计方案图面效果再理想，如果在实施过程中发生偏差，例如不按图施工、材料施工封样和设计封样不一致、施工质量粗糙等，都会影响实际建成效果。因此，项目实施过程中的现场管控，也是设计中心文旅设计部工作的重点。

2014年，文旅设计部在文旅销售物业的现场管控中投入了大量的精力。通过现场检查、现场样板段审核，设计中心对每个项目销售物业的现场进展情况、存在问题可以说是了如指掌。对于现场发现的严重影响效果品质的工程问题，一律一查到底，坚决整改，有效确保了效果品质。对于重点项目、重要节点，设计中心文旅设计部会根据具体情况组织驻场管控。如广州展示中心，在项目实施后期，设计中心文旅设计部各专业的项目负责人直接奔赴项目现场，长期驻场，协助项目公司进行现场管控，最终确保了展示中心的完美开放。

approval by specialty vice GM and department manager. All schemes, before submitted to marketing united society for approval, must be subject to approval of GM of Design Center. The rigorous requirements on scheme approval process helps to avoid different opinions of the Group leadership on the scheme, prevent the risk of later scheme rectification, ensure the scheduled advancement of project progress, and reduce the risk of the invalid cost increase.

2. ESTABLISHING GUIDING PRINCIPLES FOR PROJECT SCHEME DESIGN AND ACHIEVING UNIFIED AND CONTROLLABLE PROJECT QUALITY AND EFFECT THROUGH FACADE AND URBAN SPACE PLANNING

The Cultural Tourism Design Department initially applies facade planning for properties for sale of each project in early 2014. The facade planning realizes overall planning for space theme, facade style, block image of properties for sale concerned by proceeding from the cultural perspective, and enables the subsequent lands to be based on the unified and sustainably developed guiding principle during design, thus ensuring the coordination and integration of projects' overall effect and quality.

This facade planning work and its results are highly praised by the Group leaders. Based on specific instructions of leaders on planning work and success of facade planning, Cultural Tourism Design Department, on the basis of facade planning, proceeds with the urban space planning work, which, served as the upgrading of facade planning, can apply directional control on quality of cultural tourism properties for sale in more macroscopic, comprehensive and systematic manner. It is believed that quality effect of Cultural Tourism Properties for Sale would be further improved along with study and implementation of facade and urban space planning.

3. CONSOLIDATING ONSITE CONTROL FORCE, EMPHASIZING COMMUNICATION, TRAINING AND MANAGEMENT OF PROJECT COMPANY DESIGN DEPARTMENT AND IMPROVING ONSITE CONTROL TO ENSURE REALIZATION OF SCHEME EFFECT AND QUALITY

Given that the actual completion effect is likely to be affected if any deviation occurs to the design scheme with ideal drawing effect during its implementation, such as construction incompliant with drawings, sealed sample of material construction inconsistent with designed sealed sample, rough construction quality, etc. Thus, Design Center Cultural Tourism Design Department focuses on onsite control during project implementation as well.

In 2014, Cultural Tourism Design Department has invested a lot in onsite control of cultural tourism properties for sale. Design Center, through onsite inspection and onsite prototype section approval, is well informed of the onsite progress and available problems of properties for sale of each project. Engineering matters detected on site that seriously affect the effect quality are examined from the source and strictly rectified to effectively guarantee the effect quality. As for key projects and important nodes, the Design Center Cultural Tourism Design Department is to strengthen residence control as the case may be. Take Guangzhou Exhibition Center for example. In later project implementation period, specialty leaders of the Design Center Cultural Tourism Design Department directly headed to and long resided at the project site to assist the project company for onsite control, finally guaranteeing excellent opening of the center.

当然，真正的现场管控还是要依靠项目公司相关部门进行，设计中心加强对现场的管控，其目的不是"越俎代庖"，而是为了更及时全面地发现现场的问题。项目公司现场管控的力度、水平才是项目实施效果的最终保证。为了提升项目公司的管理，一方面设计中心加强了对项目公司的考核，督促、鼓励项目公司完善现场管理制度，提升现场管理水平；另一方面，设计中心也加强了对项目公司的培训，协助项目公司提升设计管理能力。

通过上述途径，设计中心文旅设计部在2014年，将部门的设计管控工作提升到了一个新的高度，工作成绩多次受到集团和设计中心各级领导的肯定和表扬。文旅设计部的工作也迎来了一些新的难点和压力。这些难点和压力主要来自于管控项目数量和面积的增长、品质标准的不断提升。

一方面，文旅销售物业设计管控目前从管控制度、管控流程等方面来讲，经过一年来不断地优化调整和贯彻实施，已经较为成熟稳定。难点在于不断增加的项目数量和管控面积，导致设计管控工作强度不断增加，在人员编制一定的情况下，如何确保在管控标准不降低的前提下，管控效率和管控质量的不断提升，是今后文旅销售物业设计管控的难点之一。

另一方面，万达城销售物业部分，总建筑面积动辄二三百万平方米，高的达到四五百万平方米。超大的建筑规模，意味着开发过程是一个漫长的周期，意味着万达城只能进行分期、分批开发。和所有消费品一样，客户对分期开发的销售物业也会有一个不断升级换代、品质提升的诉求，总是认为后期的产品品质应该超越前期的产品，这就对万达城销售物业的品质提出了不断提升的要求。这是今后文旅销售物业设计管控的另一个难点。

管控难点给管控工作带来压力，但是有压力才会有动力。针对上面提到的管控难点，文旅设计部也总结了未来销售物业的管控要点，包括以下几条：

第一，进一步梳理管控流程，研究将部分管控工作内容和流程分解到项目公司的可能性，这样一方面可以合理减轻设计中心的管控压力，把中心的管控精力更加有效地集中到提升项目效果和品质上，提升管理效率；另一方面，通过管控界面调整，赋予项目公司更多的管控权利，使其不再是单纯被动的执行总部的指令，可以提升项目公司的主动性、积极性。当然，这也就要求文旅设计部加强对项目公司的监管和培训，进一步提升其管理水平。

第二，加强研发工作，对项目管控流程、审核标准、设

Of course, the real onsite control still needs to be completed by related department of Project Company. The Design Center, which enhances control of the site, is not aiming to meddle in other's affairs, but to find out onsite problems timely and roundly. Onsite control power and level of Project Company is the final assurance of project implementation effect. In order to improve management of Project Company, the Design Center, on one hand, strengthens evaluation on Project Company, and supervises and encourages Project Company to improve onsite management policy and onsite control level, on the other hand, enhances training of Project Company and assists Project Company improving design management capability.

Benefiting from the aforementioned means, the Cultural Tourism Design Department of Design Center has enhanced the design control of the department to a new level in 2014 and their work achievements have frequently won affirmation and praise from leaders of Wanda Group and Design Center at all levels. Meanwhile, the Design Department is also faced with some new difficulties and pressure mainly due to ever-increasing project qualities and continuous improvement of quality standard.

On one hand, through one year's constant optimization, adjustment and implementation, the design control of cultural tourism properties for sale grows to be mature and stable in terms of control system and process. However, the ever-increasing project quantity and control area lead to escalation of design control work intensity, so the issue of how to keep improving control efficiency and quality at fixed staffing level while maintaining the control standard would be one of the difficulties that design control of Cultural Tourism Properties for Sale will face in the future.

On the other hand, as for properties for sale of Wanda City, the gross floor area may easily cover two or three million square meters or even four to five million square meters. The oversized construction scale of Wanda City naturally leads to a longer development cycle and phased development. Just like other consumer goods, the properties for sale in phased development always faces clients' upgraded and quality-improving appeal. The client always believe that products in the later period should be better than those in the earlier stage, which imposes requirements on constant quality improvement of the properties for sale of Wanda City, the other difficulty facing design control of Cultural Tourism Properties for Sale.

These difficulties will bring great pressure to the control work, however, pressure gives us impetus. To this end, the Cultural Tourism Design Department has prepared future key control points largely listed as below to cope with the above difficulties.

Firstly, to further arrange control process, studying feasibility of assigning partial control content and process to the project company, which, on one hand, relieves pressure on control borne by Design Center reasonably and enables Design Center to have effective focus on improving project effect and quality and improve management efficiency; on the other hand, through the adjustment of control interface, can grant the project company with more control power to be more active and initiative, making the company not a unit simply executing instructions of the headquarter. Consequently, Cultural Tourism Design Department thereupon needs to strengthen supervision and training on project companies and improve their management level.

Secondly, to strengthen the R&D, promoting project design

PART D EPILOGUE 后续

计导则、规划、产品、效果、品质等管控要点进行创新、标准化研发,通过这条途径,提升项目设计品质,以应对未来文旅销售物业越来越高的效率和品质要求。

第三,继续加强部门内部管理,将成熟有效的管控制度、流程固化成部门设计管理的习惯性动作,从根本上杜绝由于工作疏漏造成的设计管控失误。

在集团进行战略转型的当下,文旅销售物业扮演着越来越重要的角色,这也对文旅销售物业的设计管控提出新的要求和目标。文旅销售物业设计管控工作任重而道远,设计中心文旅设计部一定会继往开来,总结经验、提升管理,在未来的管控工作中,向集团交上一份满意的答卷!

quality to cater for increasingly improved efficiency and quality requirements on future cultural tourism properties for sale through innovative and standardized R&D against control key points, such as project control process, approval standards, design guideline, planning, product, effect and quality.

Thirdly, to reinforce internal management of departments, wiping out the design control error incurred by work omission from the root by solidifying the mature and effective control system and process as habitual behavior of department design management.

In this age of Wanda Group strategy transition, the Cultural Tourism Design Department plays a more and more important role, putting forward new requirements and goal for design control of the Cultural Tourism Properties for Sale. Design control of the Cultural Tourism Properties for Sale still has a long way to go, and the Design Center Cultural Tourism Design Department will, through summarizing experiences and improving management, keep moving forward to make Wanda Group satisfied in future design control work.

PROJECT INDEX
项目索引

WANDA
COMMERCIAL
PLANNING
2014

PROJECT INDEX
项目索引

CULTURAL TOURISM PROJECT
文旅项目

WUHAN CENTRAL CULTURE DISTRICT WANDA MANSION
武汉中央文化区万达公馆

建筑设计单位	上海联创建筑设计有限公司 北京建筑设计研究院
外幕墙设计单位	上海联创建筑设计有限公司 北京建筑设计研究院
内装设计单位	北京米多芬建筑事务所
景观设计单位	上海帕莱登建筑景观咨询有限公司 广州棕榈园林股份有限公司
夜景照明设计单位	深圳千百辉照明工程有限公司

参与人员
建筑：周群　精装：周鹏　景观：范志满　结构：田友军
弱电：宋波　强电：宋波　设备：霍雪影

HARBIN WANDA CITY EXHIBITION CENTER
哈尔滨万达城展示中心

建筑设计单位	上海力夫建筑设计有限公司
外幕墙设计单位	北京市金星卓宏幕墙工程有限公司
内装设计单位	青岛腾远设计事务所有限公司
景观设计单位	上海赛特康新景观设计咨询有限公司
夜景照明设计单位	北京栋梁照明公司

参与人员
建筑：周升森 / 邵丽　精装：薛瑜 / 陈晖　景观：薛瑜 / 杨健珊　结构：荣万斗 / 田友军　弱电：荣万斗　强电：荣万斗　设备：荣万斗

NANCHANG WANDA CITY EXHIBITION CENTER
南昌万达城展示中心

建筑设计单位	上海兴田建筑工程设计事务所
外幕墙设计单位	北京市金星卓宏幕墙工程有限公司
内装设计单位	广州思则装饰设计有限公司
景观设计单位	北京易兰建筑规划设计有限公司
夜景照明设计单位	深圳市千百辉照明工程有限公司

参与人员
建筑：周群　精装：欧阳国鹏　景观：朱作旺　结构：田友军
弱电：申茂刚　强电：申茂刚　设备：申茂刚

HEFEI WANDA CITY EXHIBITION CENTER
合肥万达城展示中心

建筑设计单位	安徽省建筑设计研究院有限责任公司
外幕墙设计单位	上海旭密林幕墙有限公司
内装设计单位	广州思则装饰有限公司
景观设计单位	上海兴田建筑工程设计事务所
夜景照明设计单位	深圳市金照明实业有限公司

参与人员
建筑：周升森 / 马申申　精装：陈晖 / 周鹏　景观：薛瑜 / 孙一琳
结构：刘征 / 邵强　弱电：薛勇 / 冯俊　强电：薛勇 / 冯俊
设备：薛勇 / 桑国安

QINGDAO ORIENTAL CINEMA EXHIBITION CENTER
青岛影都展示中心

建筑设计单位	山东同圆设计集团有限公司
外幕墙设计单位	北京市金星卓宏幕墙工程有限公司
内装设计单位	深圳市中深建装饰工程有限公司
景观设计单位	广州市东篱环境艺术有限公司
夜景照明设计单位	豪尔赛照明技术集团有限公司

参与人员
建筑：周升森 / 黄龙梓　精装：陈晖 / 周鹏　景观：薛瑜 / 孙一琳
结构：刘征 / 田友军　弱电：薛勇 / 宋波　强电：薛勇 / 宋波
设备：薛勇 / 梁国涛

WUXI WANDA CITY EXHIBITION CENTER
无锡万达城展示中心

建筑设计单位	上海中星志成建筑设计有限公司 深圳奥意建筑工程设计有限公司
外幕墙设计单位	北京和平幕墙工程有限公司
内装设计单位	深圳市中深建装饰设计工程有限公司
景观设计单位	中国建筑设计研究院
夜景照明设计单位	北京德信兄弟照明

参与人员
建筑：刘定涛　精装：欧阳国鹏　景观：杨健珊　结构：张克
弱电：宋波　强电：宋波　设备：霍雪影

GUANGZHOU WANDA CITY EXHIBITION CENTER
广州万达城展示中心

建筑设计单位	奥意建筑工程设计有限公司
外幕墙设计单位	北京和平幕墙工程有限公司
内装设计单位	广州市经艺装饰设计工程有限公司
景观设计单位	广州怡镜景观设计有限公司
夜景照明设计单位	北京三色石环境艺术有限公司

参与人员
建筑：漆国强　精装：陈晖　景观：杨健珊　结构：田友军
弱电：宋波　强电：宋波　设备：梁国涛

PROTOTYPE ROOM OF HARBIN WANDA CITY
哈尔滨万达城样板间

建筑设计单位	青岛腾远设计事务所有限公司
内装设计单位	青岛腾远设计事务所有限公司

参与人员
建筑：张天舒　精装：欧阳国鹏

PROTOTYPE ROOM OF WUHAN CENTRAL CULTURE DISTRICT
武汉中央文化区样板间

建筑设计单位	青岛腾远设计事务所有限公司
内装设计单位	北京米多芬建筑设计咨询有限公司

参与人员
建筑：周升森 / 周群　陈晖 / 周鹏

PART E PROJECT INDEX
项目索引

PROTOTYPE ROOM OF NANCHANG WANDA CITY
南昌万达城样板间

建筑设计单位 上海力夫建筑设计有限公司
内装设计单位 北京中联环建筑装饰设计有限公司

参与人员
建筑：周群　精装：欧阳国鹏

PROTOTYPE ROOM OF WUXI WANDA CITY
无锡万达城样板间

建筑设计单位 上海兴田建筑工程设计事务所
内装设计单位 青岛腾远设计事务所有限公司

参与人员
建筑：郑存耀　精装：欧阳国鹏

PROTOTYPE ROOM OF QWGDAO ORIENTAL CINEMA
青岛东方影都样板间

建筑设计单位 青岛市公用建筑设计研究院
内装设计一单位 上海腾申建筑规划设计有限公司

参与人员
建筑：周升森 / 黄龙梓　精装：陈晖 / 周鹏

PROTOTYPE ROOMS OF GUANGZHOU WANDA CITY
广州万达城样板间

建筑设计单位 广东粤建设计研究院有限公司
内装设计单位 上海腾申建筑规划设计有限公司

参与人员
建筑：漆国强　精装：李宏伟

PROTOTYPE ROOMS OF HEFEI WANDA CITY
合肥万达城样板间

建筑设计单位 安徽省建筑设计研究院有限责任公司
内装设计单位 北京东方华脉工程设计有限公司

参与人员
建筑：周升森 / 马申申　精装：陈晖 / 周鹏

DEMONSTRATION AREA OF HARBIN WANDA CITY
哈尔滨万达城实景示范区

建筑设计单位 上海力夫建筑设计有限公司
景观设计单位 上海赛特康斯景观设计咨询有限公司
夜景照明设计单位 北京栋梁照明公司

参与人员
建筑：周升森 / 邵丽　精装：薛瑜 / 杨健珊

DEMONSTRATION AREA OF NANCHANG WANDA CITY
南昌万达城实景示范区

建筑设计单位 上海兴田建筑工程设计事务所
景观设计单位 北京易兰建筑规划设计有限公司
夜景照明设计单位 深圳市千百辉照明工程有限公司

参与人员
建筑：周群　景观：朱作旺

DEMONSTRATION AREA OF WUXI WANDA CITY
无锡万达城实景示范区

建筑设计单位 上海兴田建筑工程设计事务所
景观设计单位 四川天卓设计有限公司
夜景照明设计单位 北京德信兄弟照明科技有限公司

参与人员
建筑：刘定涛　景观：杨健珊

DEMONSTRATION AREA OF QINGDAO ORIENTAL CINEMA
青岛东方影都实景示范区

建筑设计单位 青岛市公用建筑设计研究院
景观设计单位 广州科美景观规划设计有限公司
夜景照明设计单位 深圳市金达照明股份有限公司

参与人员
建筑：周升森 / 黄龙梓　景观：薛瑜 / 孙一琳

DEMONSTRATION AREA OF HEFEI WANDA CITY
合肥万达城实景示范区

建筑设计单位　　安徽省建筑设计研究院有限责任公司
景观设计单位　　上海兴田建筑工程设计事务所
夜景照明设计单位　深圳市金照明实业有限公司

参与人员
建筑：周升森／马申申　　景观：薛瑜／孙一琳

COMMERCIAL PROJECT
商业项目

NANJING JIANGNING WANDA MANSION
南京江宁万达公馆

建筑设计单位　　南京金宸建筑设计有限公司
外幕墙设计单位　上海旭密林幕墙有限公司
内装设计单位　　广州思则装饰设计有限公司
景观设计单位　　上海帕莱登建筑景观咨询有限公司
夜景照明设计单位　深圳市千百辉照明工程有限公司

参与人员
建筑：任睿　精装：彭亚飞　景观：汪新华　结构：田友军
弱电：冯俊　强电：冯俊　设备：桑国安

XI'AN DAMING PALACE WANDA MANSION
西安大明宫万达公馆

建筑设计单位　　北京中联环建文建筑设计有限公司
外幕墙设计单位　深圳蓝波绿建集团股份有限公司
内装设计单位　　北京米多芬建筑设计咨询有限公司
景观设计单位　　北京丽贝亚建筑装饰工程有限公司
夜景照明设计单位　深圳市千百辉照明工程有限公司

参与人员
建筑：石亮　精装：杨磊　景观：常春林　结构：赵可
弱电：冯俊　强电：冯俊　设备：桑国安

FUQING WANDA PALACE
福清万达华府

建筑设计单位　　福建博宇建筑设计有限公司
外幕墙设计单位　北京市金星卓宏幕墙工程有限公司
内装设计单位　　福建国广一叶装饰设计工程有限公司
景观设计单位　　四川天卓设计有限公司
夜景照明设计单位　栋梁国际照明设计中心有限公司

参与人员
建筑：孙志超　精装：胡延峰　景观：邹昊　结构：杨威
弱电：宋波　强电：宋波　设备：霍雪影

PART E PROJECT INDEX 项目索引

248
249

SALES OFFICE OF DALIAN JINGKAI WANDA PLAZA
大连经开万达广场售楼处

建筑设计单位	大连都市发展设计有限公司
外幕墙设计单位	北京市金星卓宏幕墙工程有限公司
内装设计单位	青岛腾远设计事务所有限公司
景观设计单位	大连都市发展设计有限公司
夜景照明设计单位	深圳市标美照明设计工程有限公司

参与人员
建筑：叶啸　精装：李春阳 / 刘敏　景观：张金菊
结构：李鹏　弱电：冯俊　强电：冯俊　设备：桑国安

SALES OFFICE OF PANJIN WANDA PLAZA
盘锦万达广场售楼处

建筑设计单位	北京东方华脉工程设计有限公司
外幕墙设计单位	深圳市新山幕墙技术咨询有限公司
内装设计单位	北京优яр雅装饰工程有限公司
景观设计单位	黑龙江博润景观规划设计有限公司
夜景照明设计单位	北京鱼禾光环境设计有限公司

参与人员
建筑：赵宁宁　精装：高景麟　景观：常春林　结构：李鹏
弱电：冯俊　强电：冯俊　设备：桑国安

SALES OFFICE OF DONGGUAN HUMEN WANDA PLAZA
东莞虎门万达广场售楼处

建筑设计单位	广东轻纺建筑设计院
外幕墙设计单位	北京和平幕墙工程有限公司
内装设计单位	深圳市中深装饰设计工程有限公司
景观设计单位	深圳市威瑟本景观设计有限公司
夜景照明设计单位	北京和平幕墙工程有限公司

参与人员
建筑：曹鹏　精装：梁劲　景观：邹昊　结构：黄达志
弱电：宋波　强电：宋波　设备：霍雪影

SALES OFFICE OF WANDA REIGN CHENGDU
成都万达瑞华酒店售楼处

建筑设计单位	成都基准方中建筑设计有限公司
外幕墙设计单位	成都基准方中建筑设计有限公司
内装设计单位	深圳中航装饰工程有限公司
景观设计单位	成都基准方中建筑设计有限公司
夜景照明设计单位	成都基准方中建筑设计有限公司

参与人员
建筑：赵龙　精装：宋之煜　景观：薛奇　结构：杨威
弱电：宋波　强电：宋波　设备：霍雪影

SALES OFFICE OF MEIZHOU WANDA PLAZA
梅州万达广场售楼处

建筑设计单位	广东建筑艺术设计院有限公司
外幕墙设计单位	中标建设集团有限公司
内装设计单位	上海腾申建筑规划设计有限公司
景观设计单位	深圳市卓艺泛亚设计有限公司
夜景照明设计单位	深圳市标美照明设计工程有限公司

参与人员
建筑：程征　精装：朱卓毅　景观：邹昊　结构：黄达志
弱电：宋波　强电：宋波　设备：董丽梅

SALES OFFICE OF SHIYAN WANDA PLAZA
十堰万达广场售楼处

建筑设计单位	武汉中和元创建筑设计有限公司
外幕墙设计单位	北京和平幕墙工程有限公司
内装设计单位	青岛腾远设计事务所有限公司
景观设计单位	东莞市岭南景观及市政规划设计有限公司
夜景照明设计单位	深圳金达照明有限公司

参与人员
建筑：李军　精装：杨琼　景观：高群　结构：张克
弱电：关向东　强电：关向东　设备：关向东

SALES OFFICE OF NANNING WANDA MALL
南宁万达茂售楼处

建筑设计单位	上海鼎实建筑设计有限公司
外幕墙设计单位	深圳市新山幕墙技术咨询有限公司
内装设计单位	广州燕语堂装饰有限公司
景观设计单位	北京尚艺时代建筑设计咨询有限公司
夜景照明设计单位	北京鱼禾光环境设计有限公司

参与人员
建筑：杜文天　精装：梁劲　景观：高群　结构：邰强
弱电：宋波　强电：宋波　设备：霍雪影

SALES OFFICE OF CHANGSHU WANDA PLAZA
常熟万达广场售楼处

建筑设计单位	江苏筑原建筑设计有限公司
外幕墙设计单位	北京国科天创建筑设计院有限责任公司
内装设计单位	上海腾申建筑规划设计有限公司
景观设计单位	江苏筑原建筑设计有限公司
夜景照明设计单位	常州鸿联鸿景艺术照明工程有限公司

参与人员
建筑：周恒　精装：纪文青　景观：汪新华　结构：张克
弱电：梁国涛　强电：霍雪影　设备：陈涛

PROTOTYPE ROOMS OF DONGGUAN HOUJIE WANDA PLAZA
东莞厚街万达广场样板间

建筑设计单位	广东粤建设计研究院有限公司
	广东华方工程设计有限公司
内装设计单位	广州思则装饰设计有限公司

参与人员
建筑：杜文天　精装：胡延峰

2014 WANDA COMMERCIAL PLANNING
万达商业规划——销售类物业

PROTOTYPE ROOMS OF ZHENGZHOU JINSHUI WANDA PLAZA
郑州金水万达广场
样板间

建筑设计单位　北京市建筑设计研究院
内装设计单位　广州思则装饰设计有限公司

参与人员
建筑：赵宁宁　精装：杨磊

PROTOTYPE ROOMS OF SHENYANG OLYMPIC WANDA PLAZA
沈阳奥体万达广场
样板间

建筑设计单位　大连市建筑设计研究院有限公司
内装设计单位　北京米多芬建筑设计咨询有限公司

参与人员
建筑：姚韬　精装：杨磊

PROTOTYPE ROOMS OF LONGYAN WANDA PLAZA
龙岩万达广场
样板间

建筑设计单位　中国中轻国际工程有限公司
内装设计单位　青岛腾远设计事务所有限公司

参与人员
建筑：杜文天　精装：梁勃

PROTOTYPE ROOMS OF SHUDU WANDA PLAZA
蜀都万达广场
样板间

建筑设计单位　广东蜀建建筑设计研究院有限公司
　　　　　　　北京市建筑设计研究院有限公司
内装设计单位　北京清尚环艺建筑设计院有限公司

参与人员
建筑：赵龙　精装：宋之煜

PROTOTYPE ROOM OF SHANGYU WANDA PLAZA
上虞万达广场
样板间

建筑设计单位　浙江广厦建筑设计研究有限公司
内装设计单位　上海腾申建筑规划设计有限公司

参与人员
建筑：周恒　精装：刘晓敏

PROTOTYPE ROOMS OF KUNSHAN WANDA PLAZA
昆山万达广场
样板间

建筑设计单位　江苏筑森建筑设计有限公司
内装设计单位　北京优高雅装饰工程有限公司

参与人员
建筑：刘鹏　精装：纪文青

PROTOTYPE ROOMS OF CHONGQING BA'NAN WANDA PLAZA
重庆巴南万达广场
样板间

建筑设计单位　重庆市设计院
内装设计单位　青岛腾远设计事务所有限公司

参与人员
建筑：李军　精装：董华维

PROTOTYPE ROOM OF DEZHOU WANDA PLAZA
德州万达广场
样板间

建筑设计单位　青岛北洋建筑设计有限公司
内装设计单位　广州市思哲设计院有限公司

参与人员
建筑：邵丽　精装：刘晓敏

PROTOTYPE ROOMS OF YINGKOU WANDA PLAZA
营口万达广场
样板间

建筑设计单位　哈尔滨工业大学设计院
内装设计单位　青岛腾远设计师事务所有限公司

参与人员
建筑：孙静　精装：钟山

PART E | PROJECT INDEX
项目索引

PROTOTYPE ROOM OF DONGYING WANDA PLAZA
东营万达广场样板间

建筑设计单位　悉地（北京）国际建筑设计顾问有限公司
内装设计单位　广州思则装饰设计有限公司

参与人员
建筑：邵丽　　精装：刘晓敏

DEMONSTRATION AREA OF DALIAN HIGH-TECH WANDA MANSION
大连高新万达公馆实景示范区

建筑设计单位　亚瑞（大连）建筑设计有限公司
景观设计单位　广东启源建筑工程设计院有限公司
　　　　　　　大连建筑设计研究院有限公司
夜景照明设计单位　深圳市致道思维筑景设计有限公司
　　　　　　　　　深圳市千百辉照明工程有限公司

参与人员
建筑：栾赫　　景观：张金菊

DEMONSTRATION AREA OF SHANGHAI JINSHAN WANDA PLAZA
上海金山万达广场实景示范区

建筑设计单位　中国建筑上海设计研究院
景观设计单位　深圳文科园林股份有限公司
　　　　　　　上海帕莱登建筑景观咨询
夜景照明设计单位　深圳普莱思照明设计顾问有限责任公司

参与人员
建筑：周桓　　精装：汪新华

DEMONSTRATION AREA OF SHUDU WANDA PLAZA
蜀都万达广场实景示范区

建筑设计单位　广东粤建设计研究院有限公司
　　　　　　　北京市建筑设计研究院有限公司
景观设计单位　北京易景道景观设计工程有限公司
夜景照明设计单位　北京鱼禾光环境设计有限公司

参与人员
建筑：赵龙　　精装：薛奇

DEMONSTRATION AREA OF SIPING WANDA PLAZA
四平万达广场实景示范区

建筑设计单位　北京莫克建筑规划设计咨询有限公司
景观设计单位　上海兴田建筑工程设计事务所
夜景照明设计单位　上海译格照明设计有限公司

参与人员
建筑：张爱珍　　精装：张金菊

万达商业规划
销售类物业

**WANDA COMMERCIAL PLANNING 2014
PROPERTIES FOR SALE**

2014

尹强 林树郁 曾静 张东光 门瑞冰 陈彬 王福魁 张爱珍 常春林 董莉 高景麟 李春阳 刘敏 栾赫 潘鸿岭 石亮 武宁 杨磊 张金菊 赵宁宁 昌燕 李琰 陈海燕 陈涛 范志满 李靖 李万顺 袁文卿 周妹晗 冯俊 霍雪影 梁国涛 桑国安 宋波 李鹏 邵强 田友军 张克 胡延峰 荣万斗 文善平 俞小华 马向东 叶嘯 陈文娜 刘征 薛勇 赖采靖 刘保亮 潘世伟 曹鹏 车心达 杜文天 龚芳 李军 李暄荣 姚韬 郑德延 董华维 梁勃 朱立明 高群 薛奇 邹昊 黄建好 李金桥 孙志超 黄龙梓 刘定涛 马申申 欧阳国鹏 漆国强 戚士林 孙一琳 王嵘 杨建珊 张悦 郑存耀 周鹏 周群 陈晖 薛瑜 张天舒 周升森 刘鹏 刘洋 任睿 邵丽 武春雨 张博 周恒 洪剑 纪文青 李子强 刘晓敏 彭亚飞 顾东方 申亚男 汪新华 刘大伟 钟山

图书在版编目（CIP）数据

万达商业规划 2014：销售类物业 / 万达商业地产设计中心主编．
—北京：中国建筑工业出版社，2016.2
ISBN 978-7-112-18964-9

Ⅰ．①万… Ⅱ．①万… Ⅲ．①商业区—城市规划—中国 Ⅳ．① TU984.13

中国版本图书馆 CIP 数据核字 (2016) 第 004912 号

责任编辑：徐晓飞　张　明
责任校对：刘　钰　关　健

万达商业规划 2014：销售类物业
万达商业地产设计中心　主编
*
中国建筑工业出版社出版、发行（北京西郊百万庄）
各地新华书店、建筑书店经销
北京雅昌艺术印刷有限公司制版
北京雅昌艺术印刷有限公司印刷
*

开本：787×1092毫米　1/8　印张：32　字数：850千字
2016年4月第一版　2016年4月第一次印刷
定价：750.00元

ISBN 978-7-112-18964-9
(28212)

版权所有　翻印必究
如有印装质量问题，可寄本社退换

（邮政编码 100037）